ETHICAL DIMENSIONS OF GERIATRIC CARE

PHILOSOPHY AND MEDICINE

Editors:

H. TRISTRAM ENGELHARDT, JR.

*Center for Ethics, Medicine, and Public Issues,
Baylor College of Medicine, Houston, Texas, U.S.A.*

STUART F. SPICKER

*School of Medicine, University of Connecticut Health Center,
Farmington, Connecticut, U.S.A.*

VOLUME 25

ETHICAL DIMENSIONS OF GERIATRIC CARE

Value Conflicts for the 21st Century

Edited by

STUART F. SPICKER

University of Connecticut, School of Medicine,
Farmington, Connecticut, U.S.A.

STANLEY R. INGMAN

University of Missouri – Columbia, School of Medicine,
Columbia, Missouri, U.S.A.

and

IAN R. LAWSON

University of Connecticut, School of Medicine,
Farmington, Connecticut, U.S.A.

D. REIDEL PUBLISHING COMPANY

A MEMBER OF THE KLUWER ACADEMIC PUBLISHERS GROUP

DORDRECHT / BOSTON / LANCASTER / TOKYO

Library of Congress Cataloging-in-Publication Data

Ethical dimensions of geriatric care.

(Philosophy and medicine; v. 25)
Based on the Eighteenth Trans-disciplinary Symposium on Philosophy
and Medicine, held at the University of Connecticut Health Center in
Farmington, on Sept. 20–22, 1984, sponsored by Division of Humanistic
Studies in Medicine, and Dept. of Community Medicine and Health Care in
the School of Medicine of the University of Connecticut.
Includes bibliographies and index.
1. Geriatrics—Moral and ethical aspects—Congresses. 2. Medical
ethics—Congresses. I. Spicker, Stuart F., 1937–   . II. Ingman,
Stanley R. III. Lawson Ian. IV. Trans-disciplinary Symposium on
Philosophy and Medicine (18th : 1984 : University of Connecticut Health
Center) V. University of Connecticut. Dept. of Community Medicine and
Health Care. Division of Humanistic Studies in Medicine. VI. University
of Connecticut. Dept. of Community Medicine and Health Care. VII. Series.
[DNLM: 1. Ethics, Medical—congresses. 2. Forecasting—congresses.
3. Geriatrics—congresses. 4. Social Values—congresses. W3 PH609 v. 25 /
WT 30 E84 1984]
RC925.5.E84   1987   174'.2   87–23512
ISBN-13: 978-94-010-8020-0   e-ISBN-13: 978-94-009-3391-0
DOI: 10.1007/978-94-009-3391-0

Published by D. Reidel Publishing Company,
P.O. Box 17, 3300 AA Dordrecht, Holland.

Sold and distributed in the U.S.A. and Canada
by Kluwer Academic Publishers,
101 Philip Drive, Norwell, MA 02061, U.S.A.

In all other countries, sold and distributed
by Kluwer Academic Publishers Group,
P.O. Box 322, 3300 AH Dordrecht, Holland.

TABLE OF CONTENTS

FOREWORD

There is both a timeliness and a transcendent 'rightness' in the fact that scholars, clinicians, and health professionals are beginning to examine the ethics-based components of decision making in health care of the elderly. Ethics – as the discipline concerned with right or wrong conduct and moral duty – pervades hospital rooms, nursing home corridors, physicians' offices, and the halls of Congress as decisions are made that concern the allocation of health-related services to individuals and groups in need.

In particular, care of older persons recently has received disproportionate attention in discussions of ethics and clinical care. Age alone, of course, should not generate special focus on ill individuals about whom concerns arise based on value conflicts tacitly involved in the delivery of health care. Having said that age is not the principal criterion for attention to ethics-based concerns in health care, it must be acknowledged that old people have a high prevalence of conditions that provoke interest and put them in harm's way if value conflicts are not identified and seriously addressed.

Issues that concern autonomy, the allocation of scarce resources, inter-generational competition and conflict, the withholding of treatment in treatable disease, and substitute and proxy decision making for the cognitively impaired all have special relevance for older persons. Accordingly, it is gratifying that the present volume assembles physicians, nurses, philosophers, attorneys and social scientists/gerontologists in an effort to illuminate and clarify clinical, ethics-based issues that arise in the provision of health care to older persons.

The interdisciplinary perspectives, expressed in major essays and critical commentaries of the contributors, provide a balanced and current view of the field. In addition, it is to be hoped that public and professional awareness, combined with stimuli such as this volume, will encourage investigators and funding agencies to turn their attention to clinical research in the care of the aged. Adding knowledge based on well-designed prospective studies to the plethora of theoretical deliberations accumulating in gerontology and geriatric medicine will allow decisions to be made in the future – both at the bedside and in various political arenas – that are truly moral and wise.

Travelers Center on Aging   RICHARD W. BESDINE, M.D.
University of Connecticut
  Health Center
Farmington, Connecticut

EDITORS' PREFACE

This volume is a contribution to the continuing dialogue between those disciplines that constitute the broad field of gerontology and geriatric medicine, and their intrinsically value-laden features, ethical concerns, and economic determinants. It remains to be seen what bearing the knowledge gained by these disciplines will have on the actual clinical care of the ever-growing number of aged persons who will be living as we progress toward the 21st century.

The present volume joins other volumes in this series in offering an exploration and critical analysis of concepts and values that underlie the just provision of health care for the dependent aged. In this volume, however, we consider some of the general questions occasioned by the rise of geriatric medicine – notwithstanding the fact that only at this writing and only after much prevarication has it been established for subspecialty certification in the United States.

The contributors to this volume represent the disciplines of nursing, geriatric, internal, and psychiatric medicine, medical sociology, political science, health law, journalism, public health, and moral philosophy. These multiple perspectives are required if we are to face the complex issues raised when we attend to the demands that aged dependency has and will continue to have as the actual number of aged persons rapidly increases into the 21st century. Dependency will in all likelihood have to be met by unselfish interdependency; ever-increasing average life expectancy will have to be addressed by further reliance on one another, not by selfish independency. Since it is rather unlikely that we shall ever prevent our species from growing old and dying – though the biomedical sciences might conceivably forestall the 'markers of aging,' like wrinkling of the skin or arching posture – we have already gone beyond Shakespeare's seven stages of humanity described pithily in *As You Like It*. The seventh, we recall, was mankind's "second childishness and mere oblivion, – Sans teeth, sans eyes, sans taste, sans everything" (Act II, Scene vii, 139). Thus we have already entered the eighth stage – the prosthetic age, and the seventh is no longer the "Last scene of all, – That ends this strange eventful history."

The volume which follows developed from a symposium directed to the theme: 'Geriatrics: Ethical and Economic Conflicts for the 21st Century'. This, the Eighteenth Trans-Disciplinary Symposium on Philosophy and Medicine, was held at the University of Connecticut Health Center in Farmington, on September 20, 21, and 22, 1984. During this period of gestation, much has been altered and some additions have been made. Its final form emerged from discussions between the participants and the authors of papers and commentaries. We acknowledge our appreciation to the many individuals who attended this symposium and who, through comments and discussions with the authors, enabled them to revise and frame their essays as they now appear in this volume. We wish, as well, to thank all of those who have aided us in developing this volume from the proceedings of the symposium. From among the many to whom we are in debt, we would especially like to mention Susan G. Engelhardt, Series Editorial Assistant, who carefully reviewed the final manuscript and page proofs; in addition, Sarah K. Beckett, J.D., offered numerous comments and suggestions during the time the manuscript was in final preparation; Veronique Ingman, who kindly prepared the Index; and Ms. Theo Ungewitter, who inherited the post-symposium manuscripts, unselfishly re-typed the manuscripts and incorporated the numerous revisions.

We also wish to express our gratitude to all who helped to make possible the symposium from which this volume developed. We are grateful to the Division of Humanistic Studies in Medicine and the entire Department of Community Medicine and Health Care in the School of Medicine of the University of Connecticut, who sponsored the program, and especially to Cecile J. Volpi, Head, Office of Continuing Education, who with her able staff made all of the local arrangements which enabled the participants and the public to feel welcome during their time at Farmington.

The project was supported in part by a grant from the Connecticut Humanities Council, the State Committee of the National Endowment for the Humanities; however, the viewpoints or recommendations expressed in the volume are not necessarily those of the Council or the Endowment. The editors wish to express their gratitude to the Council and its Executive Director, Bruce Fraser, for their support of the symposium. In addition, Connecticut Long-Term Care Research, Inc., through Joan Quinn of Connecticut Community Care, Inc., and the University of Connecticut Research Foundation, generously provided additional material support which enabled the symposium directors to execute their original program without revision. We stand in special debt to the University of Connecticut Health Center's administration for the ways it encouraged and made possible this meeting of scholars and the public. Many individuals labored unselfishly in the preparation and conduct of this symposium on philosophy and medicine. We are deeply grateful to them all; through them, a symposium came into being that allowed sustained cross-disciplinary discussion concerning the inter-relations of the major secular learned professions with respect to the problems of providing health care to the aged.

After two years of development from the proceedings of this sympo-sium to the volume, a set of sustained analyses has been forged. As a series of interwoven discussions in print, it ranges in focus from public health issues to the basic ethical and legal underpinnings of the care of the aged infirm by the newly-burgeoning clinical domain of geriatrics. This volume, however, makes no claim to encompass all of the relevant issues, or even all of the central issues at stake. However, like earlier volumes in the *Philosophy and Medicine* series it is offered as an introduction to some of the central conceptual aspects of the biology and epidemiology of aging, the equitable provision of the medical care for the dependent aged, the meaning of self-determination in later life, and the numerous issues that bear on the just provision of health care for the aged.

May 31, 1987   STUART F. SPICKER
                                                      STANLEY R. INGMAN
                                                      IAN R. LAWSON

INTRODUCTION

The population of the United States, as in other developed countries, has aged dramatically during the present century. Americans sixty-five years of age and older, who accounted for only 4% of the population in 1900, represented more than 11% of the U.S. population by 1980. This trend toward population aging is expected to continue well into the 21st century. The contributors to this volume explore the value conflicts likely to surround continued aging of the U.S. population: in regard to the health and social status of the elderly of the future, their potential for self-determination in later life, and their requirements for health care.

The central theme of the volume is that continued growth of the aged segment of the population, a quantitative phenomenon, will inevitably pose challenges of a normative and ethical nature to the old and young alike. These challenges may center on coping with the increased dependency of advanced age, on developing the fiscal and organizational resources to deliver currently available health and social services to ever greater numbers of elderly, and on equitably reformulating our social responsibilities to the aged. At the same time, effort may be required to preserve a vision of aging which encompasses the healthy as well as the dependent elderly, and to carve out meaningful roles for those of advanced age.

From now until the middle of the next century, it appears highly likely that the percentage of the population that the aged constitute will increase. As a consequence, the number of individuals suffering from serious physical and mental disabilities such as heart disease, cancer, arthritis, Alzheimer's disease, and various forms of senile dementia will grow larger. These projected increases have dramatic implications for rising health care costs just as public policy strategies are being developed to contain such dramatic escalation of the costs. New conflicts are likely to emerge between macro-allocational choices made within commitments to cost containment and clinical micro-concerns in providing elderly individuals with needed health care, especially when those individuals suffer from severe mental and physical impairments.

Traditionally, physicians have tended to treat the elderly less aggressively than those in the prime of their lives. In addition, circumstances
such as the physical and mental impairments of elderly patients have
weighed heavily in physicians' decisions not to pursue extensive or
costly attempts to treat. To be sure, such choices are often couched in
terms of medical indications taken to reflect the decreased likelihood of
benefit to persons from major invasive procedures when they are
advanced in age and have serious chronic diseases or disabilities. Such
decisions may, though, also mask social and ethical judgments regarding
the quality of life of the individuals involved. Age alone, for instance,
does not appear in some settings to be a contraindication for the use of
the intensive care unit, though elderly patients appear to have a high
mortality rate during the year following discharge. It is, however,
plausible that tendencies to provide less intensive or less costly care to
the elderly will be augmented, given increased interest in containing
health care costs (e.g., through prospective reimbursement plans for
Medicare payments to hospitals).

Whatever the precise nature of the challenges ahead, our current
understandings of demography, and of the relationship between health
and aging, suggest that they will be formidable, and that the concerns
expressed by the contributors to this volume are vital.

Current Bureau of the Census population projections show the number of Americans aged sixty-five and older increasing from 25.7 million
in 1980 to 35.0 million by the year 2000, and to 67.1 million by 2050.
Should these projections prove accurate, the elderly, who constituted
11.3% of the population in 1980, will account for 13.1% of the population in the year 2000, and represent 21.7% of all Americans by the
midpoint of the 21st century [1].

Are such projections, which partially reflect declining mortality rates
at all ages over the last decade and a half, plausible? Demographers
Vaupel and Gowan have shown that, assuming constant fertility rates
and zero net migration, the proportion of Americans aged sixty-five and
older, in the absence of any further reductions in mortality rates, will
still reach nearly 20% by the year 2080. Under similar conditions, a 50%
reduction in mortality rates at all ages, achieved in the year 2000, would
result in 27% of the population being age sixty-five and older in 2080.
Finally, under a scenario involving a steady reduction in mortality rates
of 2% per year at all ages, approximately 38% of the U.S. population
would be at least sixty-five years of age in 2080, with nearly 19 million

Americans aged 100 or older. While increases in fertility or net immigration to the United States might moderate any of these outcomes slightly, the overall trend toward a significantly older population would not substantially change. Moreover, selective reductions in mortality favoring the elderly could just as easily accentuate their numbers in the future [10].

Despite declining mortality rates among the old and very old, Americans aged sixty-five and older are at greater risk of serious illness and its complications than younger age groups. In particular, they are more susceptible to chronic disease and long-term functional limitations. This susceptibility is reflected in the extensive health services utilization of the elderly. They see the doctor more often, make greater use of ancillary health services, account for a disproportionate share of prescribed medications, are more likely to undergo hospitalization, and occupy the vast majority of nursing home beds in the U.S. [6]. As a result, their per capita health care expenditures are seven times greater than Americans under twenty years of age, and two and one-half times greater than Americans aged twenty to sixty-four ([4], p. 152).

Furthermore, the elderly's health problems are often compounded by social and economic problems. As a group, for example, they are at greater risk of social isolation, with nearly one out of three elderly in the community living alone in 1984 [5]. While the proportion of elderly living below the official poverty level has declined in recent years, fixed incomes and a heavy reliance on social security benefits are common, with substantial numbers of elderly at or marginally above the poverty line ([9], p. 43).

Rice and Feldman have projected the impact of continued aging of the U.S. population in light of the elderly's more frequent morbidity and greater use of health services [7]. In their analysis they apply age and sex specific disability and health services utilization rates from the late 1970s and early 1980s to Social Security Administration population projections through the year 2040. Their quantitative projections will disturb many health and social service providers, and those responsible for the formulation of public policy with respect to the aged. If the present age-sex specific rates of functional limitation remain constant through the first half of the 21st century, the number of non-institutionalized Americans requiring assistance with one or more activities of daily living (ADL) such as walking, bathing, using the toilet, dressing, eating, or getting in and out of bed, will increase by 152%

(from 3.1 million in 1980 to 7.9 million in 2040). Similarly, the number of physician visits annually will increase by 47% (from 1.1 billion to 1.6 billion), the number of short-stay hospital days by 100% (from 275 million to 549 million), the number of persons in nursing homes by 246% (from 1.5 million to 5.2 million), and annual expenditures in constant dollars for personal health care by 68% (from $219 billion to $369 billion). Approximately 87% of the projected increase in ADL-limited persons, 50% of the increase in physician visits, 76% of the increase in short-stay hospital days, 99% of the increase in nursing home residents, and 69% of the increase in health care expenditures will be due to an increase in the sixty-five years and older segment of the U.S. population.

Should they occur, these increases in health services utilization will take place during a period of change in the nature of support ratios in the U.S. For example, as growth of the aged segment of the population exceeds anticipated declines in the proportion of persons under twenty years of age, there will be a reduction in the proportion of working-age Americans. Recent Bureau of the Census population projections suggest that the proportion of 20 to 64 year olds will drop from 58% in 1982 to about 55% in 2050, with the 'total' dependency ratio in the U.S. increasing from approximately 731 persons of 'non-working' age for every 1000 persons between 20 and 64 in 1982 to about 820 in 2050 [1].

Overall, the economic impact of this change on the Americans of working age could be slight. Since young people receive considerably more support than older individuals when both public and private resources are considered, the relative cost of 'supporting' younger and older citizens could even decline. Given the elderly's greater reliance on public support ([8], p. 24), however, it is probable that meeting their needs in the future will require the transfer of surplus private dollars – no longer needed for the declining number of young Americans – to public programs for the growing number of aged Americans. Such a transfer may be difficult to accomplish in a social and political environment where 'cost containment' is stressed and where the needs of the elderly must compete with other interests for available public dollars.

While projected increases in the proportion of aged Americans have generally prompted concern on the part of health and social service providers, program planners, and policy makers, important questions have been raised regarding the actual impact of population aging on the well-being of the elderly and their need for organized medical, health, and social services. Critics have asked, for example, whether it is appropriate to project the needs of elderly in the future on the basis of what we know about today's elderly. The number of elderly will certainly increase in the 21st century, but will their health and social needs remain the same? In particular, could the factors contributing to reductions in mortality in old age also lead to improvements in health status in old age?

J. F. Fries, a physician, drawing upon the observed reductions in mortality, accompanying increases in average life expectancy, the assumption of a relatively fixed life span, and an epidemiological model of disease, has constructed a seductive picture of a future in which serious illness is 'compressed' into the final years of old age [2]. In his analysis, Fries attributes the twenty-six-year increase in average life expectancy experienced in the U.S. during this century to improving environmental and social conditions, the advent of effective public health measures, advances in health care delivery, and related societal changes. These improvements prolong life by reducing the risk of contracting many serious diseases, and by minimizing the consequences of such diseases when they do occur.

Carried only to this point, Fries' analysis helps to explain the concentration of chronic disease and disability in old age, but provides little optimism for improvement in the health of the elderly in the future. Fries, however, carries his analysis a step further. Asserting that the human *life span* (longevity) is biologically fixed at 110–115 years, he proposes a practical limit to the *average life expectancy* attainable in any human population. Since no population will be entirely successful in eliminating either disease or trauma-induced mortality, this limit must be significantly lower than the human life span. Based on historical changes in survival curves, Fries sets this limit at approximately 85 years in developed societies. With an average life expectancy at birth of 78 years for women, the U.S. is fast approaching the practical limit of life expectancy. Under these circumstances, the impact of past and future improvements in living conditions and health care could be to push the age of onset for many chronic diseases (or their clinical manifestations) toward or beyond the average life expectancy. The result would be a reduction in the level of chronic illness and associated disability in old age, their compression into the final few years or months of life, and an increase in the incidence of 'natural death' unaccompanied by extended illness. Such improvements could significantly offset the increase in health care resources which would otherwise be required to meet the needs of a larger population of elderly.

In the essays that follow there are concerns about dependency burden, cost containment, and the efficiency and distribution of material resources, all of which serve as stimuli for economic conflicts. The term 'ethics' is employed in this volume not in the narrow sense of rationally justifying individual conduct, but to suggest problems of social and/or distributive justice, as well as issues which call attention to the rights and responsibilities both of individuals and collectivities. To address these issues the volume has been structured in four sections.

Initially, the contributors to Section I work to define current and future needs in geriatric care, i.e., what are or will be the likely demands upon the collectivity, or upon individuals and their immediate kin and friends. In Section II, the contributors outline some of the weaknesses and strengths of the present geriatric health care context in the U.S. What emerges is a continuous debate about where the principal responsibility should rest. The key question is asked: Which policies are the most effective and equitable? Section III focuses more deliberately upon the micro level: the providers who interact with others who may or may not be patients. In Section IV the essays generally reflect two opposing views of the world: one encourages more individual responsibility; the other calls for more collectivist approaches. Both seek to bring about greater justice. Finally, the epilogue offers a vision for the future – the eighth stage of humanity – the 'prosthetic society', a society in which a greater effort is undertaken to accommodate the disability, disease, and dependency that accompany old age.

Opening Section I, Jacob A. Brody, a physician and epidemiologist, asks the rhetorical question – "Is this the best of times or the worst of times?" – with respect to aging and dependency. He challenges the now famous arguments proffered by Dr. Fries. Brody asks us to ". . . focus on the realities of the onrushing future"; he persuasively challenges Dr. Fries' and others' tantalizing visions of old age devoid of significant infirmity and dependency. He points to the long-term nature of declining mortality and increasing average life expectancy, to the human body's increasing ability to outlive its parts, and to the sobering implications of these two facts when considered together.

According to Brody, "we simply do not know the limits of life expectancy," but with average life expectancy increasing faster for the aged than other age groups, disability will increase in significance. The key is to plan to postpone, eliminate, or mitigate the burdens of disability. For Brody this is a difficult but not a bleak project. In short, we will be obliged to devote a greater proportion of our resources – material and human – to restorative procedures, and we will also need to focus attention on the complex bioethical issues relating to the ways we die. Ultimately, Brody's careful review of the demography of aging leaves us with little hope that the 'hard choices' discussed by the volume's remaining contributors can be avoided.

Tom Beauchamp, a philosopher, returns to basic questions in his response to Brody's essay. Virtually all of Brody's major claims are challenged, and Beauchamp asks on what basis can one assume that past responsibilities or rights will continue into the future. For example, what are the grounds for claims like, "We will be obliged to devote more resources to the aged"? Beauchamp's many questions probe the limits of our responsibility and the limits of a government's responsibilities to its citizens. He focusses on Brody's projections – presuming they are correct – and asks what will be the "unbearable social costs". He also presses for us to outline a system for equitably rationing health care services, and thus to rationally determine what services will be given to all, and which to only a few. Beauchamp desires a health care system that will be compatible with the demands of justice, and wonders whether a cost/benefit analysis should be used to establish such a system. In short, Beauchamp, like H. T. Engelhardt, Jr., later in the volume, asks to what degree one must provide resources for the health care of the aged. Brody apparently by-passes such questions and hence Beauchamp asks: "Is bad Samaritanism unjust?" In the end, Beauchamp offers an illustrative example to show that the trade-off between encouraging older faculty members not to retire (based upon new life expectancy projections) and the rights of younger faculty in need of academic positions is intractable conflicts that have staggering implications. Is Beauchamp too pessimistic about the nature of our collective ability to restructure social organizations and systems?

Albert Rosenfeld, a journalist, argues that today's prevailing image of the aged is one of dependency. He suggests that numerous writers tend to encourage this image of dependency, which he believes reinforces old people in their thinking that they are incompetent, or that they should not try to remain independent as long as possible. Thus Rosenfeld calls for a reappraisal, noting that the present image emerges from two essential facts: (1) the ever-increasing number of aged persons and (2) the increasing average life expectancy and longevity or life span of our species. This translates into a powerful negative image: an increasing number of infirm and burdensome aged. But if we overestimate aged dependency, the projections for the future will not be realistic. Thus Rosenfeld shifts the discussion from the 'burden' of the aged to the ways in which the young may eventually relate to the energetic and independent elderly. His solution is to give attention to *interdependency*. Thus he is optimistic with regard to the avant-garde gerontologists who, along with a number of geriatricians, call for radically different approaches in the clinical care of the elderly. At this point Rosenfeld rehearses a number of scientific theories of aging – cross linkage, free radicals, somatic mutation, decline in immunity and autoimmunity – and he suggests that when we discover the secret for inhibiting aging, we can then actually prevent aging and go beyond or by-pass the prosthetic age. Problems arise, however, in Rosenfeld's casual use of the term 'aging'. For example, in such remarks as – "As I age my skin becomes wrinkled" – aging is the *cause* of wrinkling. On the other hand, "My body's rapid changes (in physiology) cause me to age quickly" suggests that bodily changes *cause* aging, and aging is the result of internal and external organic decline. The obvious circularity here suggests that 'aging' is an unclear concept in need of clarification.

It is important to note that writers like Rosenfeld take the *markers* of time passing (age?) as the proper meaning of aging, and aging is thus not a psycho-historical phenomenon. Another way is to suggest that we refine our use of terms like 'aging' and 'aged' as well as 'aging slowly' and 'aging rapidly'. Thus, cells do not age, only persons do. In short, contrary to Rosenfeld's view, one can argue that biologists do not study the mechanisms of human aging, but rather the changes at the cellular and sub-cellular levels which occur over time. 'Aging' should properly be used to describe personal biography in the penultimate years. We are easily misled when 'aging' is employed to refer to biological processes producing the *markers* of time. Human aging is essentially historical, experiential, and biographical. Careful biological investigations into the changes that occur in the cells and organ systems of the human body are not studies of aging at all; one must therefore read Rosenfeld's essay with care. In sum, biologists are not searching to uncover the structure of cellular and sub-cellular processes in order, in time, to discover how to keep young persons from becoming old persons (as Rosenfeld might have us believe), although they are interested in discovering the mechanisms by means of which, if controllable or reversible, they could conceivably enable a 90-year-old person to *look like* a 50-year-old person typically looks today.

Randolph Nesse, a psychiatrist and biologist, offers a particular theory of aging so that it is conceivable that one day a means for preventing the markers of aging will be available. Nesse appeals to an evolutionary theory of aging. Is he, like Rosenfeld, Fries, and others, hoping for the fate of the "one hoss shay" as the instantaneous end of life without illness, disability, and dependency? Nesse's attention to an evolutionary theory of aging is, of course, linked to biological science, and hence he explores the relationships that are possible to account for aging in terms of Darwin's theory of natural selection, and the knowledge that has been acquired since Darwin's time. Nesse provides a very original view, one which takes into acount advances in our understanding of genetic mechanisms and the genetic heritability of traits, as well as the transmission of those traits that are functionally adaptive for the aged members of our species.

Daniel Hausman, a philosopher, introduces Section II by turning sharply to the new for-profit market approach to medical care and offers an explanation of the need to modify it to meet the needs of the aged. For him, individuals are acting imprudently if they attempt to make provision for their later life by some form of "heroic savings." The purchase of health insurance is what the "ideal rational economic man" does. But what kind of insurance is appropriate? Because Hausman views medical and health care as basic needs, he argues that the state must provide a system to protect the indigent. The critical question remains: Is the market approach the best method for lowering expenditures? In his review of the chief arguments which bear on the market model, Hausman clarifies both the strengths and weaknesses of this view. In short, for Hausman health care is a basic need and thus poverty should not prevent anyone from receiving competent health care. However, the goal should be a decent minimum level of care. Even the British National Health System, he reminds us, allows the rich to purchase what they deem to be "better health care." To eliminate this privilege would cause too much oppression and is not worth advocating. Analyzing this, Hausman attends to the question of society's versus the individual's responsibility for ameliorating health problems. For him, employers who increase health risks for others, as well as citizens who take known risks, must themselves pay for the resulting costs. However,

Hausman does not advocate a regulatory mechanism for those persons who practice unhealthy life styles. In short, collective mechanisms are superior to market mechanisms in terms of the equitable distribution of expensive medical technologies and treatments; thus we cannot, he argues, simply hide behind market forces; to do so betrays moral cowardice.

Baruch Brody, a philosopher, reviews the recent economic situation, e.g., Medicare going broke and the Medicaid crisis. Why? Look to the (1) increase of general price inflation, (2) more visits to physicians or hospitalizations per capita, and (3) more services per visit.

Brody's basic claim is that "most of the inflation adjusted growth is due to good goals, not to waste." He then considers the rising percentage of GNP being consumed, and asks: What is the likely impact on the aged? Specifically, how would a regulatory versus a market approach to cost containment affect the aged? Both might reduce waste but their impact is different when rationing is the issue. He reviews current cost control systems: (1) peer review and (2) prospective reimbursement. He explores these as strategies affecting a national rationing policy. But, he asks, "How shall we ration?" He considers three approaches: (1) a national medical policy which sets specific limits, (2) a national budgetary policy which leaves decisions to individual providers, and (3) a national social policy which increases patient participation. He supports (3) yet considers the problems with this approach: (A) providers are not comfortable with it, (B) health care recipients and/or their families will not accept it, and (C) there is no social consensus about the type of cases to which the model would apply.

Thomas Halper, a political scientist, notes confusion in the issue: Since medicine shapes our outlook of ourselves, our society, and our future, greater resources of all types go to medicine and health care than to national defense. Halper agrees with Brody that our society's commitment to provide care for all is *the* major cause of increased costs. He asks for more reflection on Brody's arguments concerning inflation in the health field. First, should inflation really *not* be regretted? After all, vast sums are spent on the last year of life. In addition, Halper thinks Brody should address the means of providing services, e.g., the entitlement principle versus the need principle. Is it not more just to redistribute resources in the interest of the less well off than simply in the interest of the sick, the injured, or the aged? Since the aged exceed the national average in net income and total assets, transfer of funds to them through Medicare may not be equitable.

Halper then turns to Brody's assumption that the aged will overburden the system. Victor Fuchs, he argues, showed that the anticipated burden is really overestimated. Lastly, Halper asks if rationing is necessary, as Brody has argued. After all, rationing is not a new notion, only more visible today. In the end, Halper is confident that the U.S. will adjust to cost effective strategies since our history suggests this is how we typically face the future.

T. Forcht Dagi, a neurosurgeon, recalls that as a senior registrar he noted the aged were accorded a different standard of care, e.g., limited access to hospitalization, intensive care, and, ultimately, resuscitation. A form of cost-benefit analysis was used to justify such decisions. Although this was some years ago, and even though there was a negative reaction to such an approach, this tendency still prevails.

First, Dagi deals with the various historical definitions of terms like 'resurrection', 'revival', and 'resuscitation'. He then provides a brief account of the differences between them. Under presumption of life or death, death threatens in revival, and has occurred prior to resurrection; one revives when "capable of maintaining vital signs prior to the patient's revival;" but revival is the weakest form of intervention; in terms of outcome, success is expected in revival, reasonably anticipated in resuscitation, and miraculous in resurrection. Resuscitation is required when an irreversible, underlying condition is responsible for an acute event. With revival, the patient is self-sustaining, whereas in resuscitation he may not be self-sustaining. Resuscitation done routinely in emergency rooms on the clearly dead, then, should properly be called 'resurrections'. Dagi turns to the Moral Imperative: revive the faint and the fallen; for those trained there is an obligation to resuscitate.

At this point Dagi introduces a new term for interventions – 'reconcinnation' – from the Latin *re concinnare* which means to restore, reconstruct, mend or refit the thing harmoniously. Dagi considers in depth "the obligation to resuscitate." Historically, the success of cardiac massage for cardiac arrest reflects the belief that the restoration of life should be attempted. The difficulty of distinguishing between death and the appearance of death also supports intervention. By the end of the nineteenth century, a great deal of energy had been expended to return the dead to life by way of technical procedures. The aging factor also affected the "obligation to resuscitate." There are two conflicting tendencies: (1) the aged were excluded from the highest priority of care; (2) the aged were respected! More recently the slogan, "death with dignity," is pitted against the desire to improve geriatric care. There are two reasons for this: (1) being overwhelmed by the burden of the patient's age the physician tends to forget the importance of restoring function; (2) physicians can misinterpret the maxim – "if you cannot care, first do not harm" – as "do nothing if you cannot cure."

Finally, Dagi turns to "do not resuscitate" orders (DNR). Technology and various methods to determine a patient's wishes have served to change today's situation. The extant literature on quality of life and cost-benefit issues reflect emerging policy positions. Dagi concludes that DNR decisions should be related to a theory of distributive justice, since such decisions may cause damage to both patients and society. He concludes with six principles that bear on the writing of DNR orders.

Molly Gavin and Gail Kataja describe two "high risk, frail elderly" cases to introduce Section III, cases which they assessed and for which they coordinated all services, benefits, and entitlement as well as monitored "the quality of life" of the patients in response to intervention. While these cases do not reflect the independent, prudent consumer of medical care as idealized by Daniels, Beauchamp, and Halper, these nurse clinicians (as well as their colleagues in social work) must determine daily how far to go in making or influencing decisions for their patients. They realize that objective assessment is difficult. They note, for example, that nursing home assessors are more likely to recommend nursing home placement. Would home care assessors who work for hospital-based home care programs be more likely to recommend hospital admission than those who work outside of hospitals?

These two cases and questions set the stage for Nancy Dubler and her discussion of the legal rights of the dependent elderly. According to Dubler, the range of choices tends to enhance the quality of life in American society. Most choices, however, are unregulated by society. Yet, dependent elderly persons are often denied the right of free choice. What are the legal issues? First, caregivers should document the individual's desires and design plans to maximize autonomy, limit paternalism, plan for cognitive decline situations, consider the "best interest" of comatose patients, and, finally, "fight against defensive medicine." The term 'elderly', she reminds us, does not always signal 'dependent' in discussions of autonomy and competence.

After an extensive discussion of the concepts of competence and autonomy, Dubler suggests that there is need of a new legal doctrine. This doctrine should revolve around "supported judgment," that is, what is this person's "spoken choice?" What is his or her present articulated preference and does it relate to prior patterns of preference? She urges us to remember that forced dependency or diminished autonomy lowers perceived competence; or, more generally, competence levels are societal artifacts – a truly socio-psychological phenomena. With demented patients, their prior strong preference should not be overriden easily.

Dubler then considers the role of proxy decision makers, that is, conservators and guardians. How can abuse be avoided? Does guardianship lead to categorical infantilization? Surprisingly, few elderly and few counselors for the aged are present at court competency hearings. These "nonadversarial" proceedings in particular operate with little understanding of gerontological science. A new elder-court system could, theoretically, help avoid some of the weaknesses of the present system. An interdisciplinary consultation should be established for certain types of dependent geriatric patients; in addition, in a few rare cases, an elder court of appeal should also be formed to deal with a patient's adamant refusal to comply.

In her essay, Margaret Battin, a philosopher, notes that dependency awaits many of us, and therefore suicide may become an attractive option for some. She cites Robert Kastenbaum, who argues that suicide can be expected to become the preferred mode of death, because by electing suicide one maintains control over all relevant aspects of one's death. Battin examines this thesis by considering two major questions: (1) Is it rational? (2) Is it morally justified? One's emotional reaction to these arguments is similar to the reactions of the Gray power group, when they denounce stereotypical forecasts of extreme dependency and powerlessness; they warn that uncritical acceptance of these predictions invites the very conditions it forecasts. Such "arguments" allow us too easily to raise the suicide option to the level of social policy.

Battin outlines the reasons she views suicide as a poor option: it precludes engaging in enjoyable activities; it interrupts ongoing projects; precludes future sensory experience and pleasures, and human communication; it damages one's reputation; it risks severe pain and, for some, prevents attainment of salvation after death. After evaluating these reasons, she argues that they may only have limited relevance for persons of extreme old age. In his commentary, Joseph M. Healey, J.D., closes Section III by recapitulating the points of commonality in the three previous articles, and emphasizes some important points concerning the legal lattice which surrounds the intervention or intrusion of medical service providers into the lives of elderly patients. Problems still remain, e.g., how to justly distribute health care resources for the aged; how to make caregivers into "justice officers" when mercy and justice are in conflict.

But what does justice require in the distribution and provision of health care resources?

Section IV is introduced by Norman Daniels, a philosopher, who views the health care system as a "saving scheme," a deferred use of resources from one stage of life to another. A prudent design would recognize the "saving scheme" and rationing would be justified. But what is the proper relationship between the entitlements of the elderly and our prudential thinking on what the system should do?

Daniel's general theme is as follows: We want institutions to be responsive to us at all stages of life and for all medical needs; thus, if we use resources too early, perhaps we suffer later, because institutions at some stage of life will not have the services. Thus budgeting is unavoidable. Waste and inefficiencies in resource use do not allow us to avoid rationing. Moreover, we should attempt to avoid the situation where the young are pitted against the aged, that is, where each takes resources from the other. Instead, Daniels views it as an individual problem – "you only see yourself at other stages of your life benefitting from your own prudent savings." Individualism is to be preferred for it avoids the "system that takes from some to make society as a whole better off." Such systems, he feels, risk or fail to respect the importance of the person. Daniels goes on to consider when rationing by age is appropriate for life-extending resources. His approach "avoids much that is controversy about rationing," and he argues for individuals to construct their own benefit packages. Then he reviews most carefully the use of age as a criterion for rationing life-extending resources. He thinks his "prudential model" is less controversial, for his "insurance package" is intended to resemble some features of the British National Health Service. Like Ingman, Gill, and Campbell, he points to limited hemodialysis care in Britain where home care is more extensive. How would it operate? Could each individual select his own rationing system? Could each change his mind at a later time?

In sum, Daniels argues that health care needs are of special moral importance. His scheme, he believes, protects an individual's fair share of the normal opportunity range. But rationing is still a critical concern. He discusses two rationing schemes: In Scheme A (Age Rationing) no one over age 70 or 75 is eligible. Scheme L uses Lottery Rationing. Daniels argues for Scheme A, since late life is typically less valuable to us. He argues that his position may help us evaluate the British system of rationing renal hemodialysis by age, for it presents the conditions in terms of which to judge the justice of their system. In the end, he urges us to learn to think prudently and prospectively about the kind of health care we prefer. Physicians should also think about the "gate keeping" decisions that they make, and they should be persuaded to avoid under-the-table forms of rationing by age. Instead, collective or social decisions should be used to guide providers; only then can we avoid the split between "them" and "us." After all, they are decisions about ourselves.

Ingman, Gill, and Campbell, medical sociologists, examine the delivery of geriatric care and renal care in the U.S., U.K., and Switzerland. They are primarily concerned with the ways these three nations' macro-allocation policies differ and what the consequences are for individuals in these societies.

They select two areas of medical care provisions – renal care and geriatric care – which have different social and historical characteristics. Geriatric care has an obvious social dimension; renal care is strongly linked to high technology medicine. They attempt to categorize two models – either "individual" (market oriented) or "collectivist" (social-ist) oriented. The United Kingdom, with its National Health Service, is a collectivist-oriented system; whereas Switzerland, closer to the U.S. tradition, stresses individualism and a market-oriented health care policy.

Relative to the standard of living and the available resources in these societies, the authors believe that the evidence reveals that the U.K. aged secure a more equitable distribution of available care than either the aged in the U.S. or in Switzerland. With respect to renal care, renal-failure patients in the U.K. are more often denied renal dialysis than their counterparts in Switzerland and the U.S. Thus a national health system, which initially receives high marks for equity and uniformity, breaches these principles with regard to access to renal dialysis. The authors then suggest that there are more direct political mechanisms available to correct these apparent injustices, at least for other nations than the U.S.

Engelhardt, in his commentary on Daniels and Ingman *et al.*, explores the limits of public authority to redistribute wealth and individual energies. As he argues, if the state does not own all of the possessions and services of its citizens, then individuals will always have a right to trade freely at least some of their resources for the services of others. If that is so, individuals will always have the moral right to buy around government systems attempting to establish a single system of health care distribution. Moreover, attempts to fashion a single inclusive system of health care distribution must assume that there is a uniform or morally canonical view of how resources should be applied – and such does not appear to be discoverable. Provision for the elderly must be understood not in terms of appeals to discoverable *a priori* patterns of justice, but through the model of insurance programs created publicly and privately in order to realize different views of how to provide health care. Because of the limits of state authority and because of the divergent views of how to come to terms with aging and health care needs, inequalities and differences in health care provision will not only be practically but morally unavoidable.

Some thirty years ago, Scotland's medical care system transitioned to a newly created geriatric medical and social care service within a new national health service, a system with limited capital resources in terms both of long term care beds or hospital beds, a prospectively budgeted specialty-hospital care network with its salaried physicians and per-capita system in which primary care physicians were paid so much per patient per year. In the geriatric care service, newly appointed consultants or geriatricians were asked to be "prudent consumers" and ration resources for all patients referred to them by general practitioners. "Third parties" were the regional hospital authorities, with the national ministry of health possessing overview and resource allocation power. Insurance, public or private, was eliminated for the most part – except for the very wealthy. Rationing for the aged was embodied in the "pre-admission assessment process" controlled by the geriatrician – the clinical-manager or gatekeeper for expensive medical care services.

Based on his "bio-sociology," Dr. Lawson outlines in a personal Epilogue the main feature of what he calls a "prosthetic society." This society is designed around three dimensions: disease, disability, and dependency. It is a system which is friendly – much like the modern

personal computer – but far more complex. Some things seem simple
(e.g., house calls, ramps to offices, receptionists who organize transport), but other items require macro-system reform (e.g., interconnect-
edness of component services to assure continuity). He worries that the
corporatization of American health care is not likely to encourage
empathy, altruism, advocacy, or hard work; it is not user friendly, e.g.,
Medicare and Medicaid create fiscal barriers like "deducts" and "copays"
that encourage decreased use. At the same time, the current system
apparently supports the principle that "megabucks" should be made by
the private sector or things won't happen right.

Of course, our 1985 "futurism" in ethics and economics, applied to
the elderly, has been overtaken by more recent events. Prospective
payment systems, e.g., Diagnosis-Related Groups, Lawson holds, have
so rationed hospital care to the elderly that, by common consensus, they
are at times being discharged quicker and sicker. But more severe fiscal
restrictions are also being applied to them when they are dispatched to
home care and nursing home care.

The Institute of Medicine has recently published a report on the
rather unsatisfactory state of the nation's nursing homes and has recommended neither better doctoring and nursing nor more money, but an
even tighter structuring of existing regulation by government [3]. The
American Medical Association has recommended alteration of the
Medicare system because of the alleged inequity involved when younger
workers will have to support the old in medical need. It is conceived that
we all, by prudent thought and saving, can take care of our own selves
personally in the unimaginable vicissitudes of 30 and 40 years hence.
But Lawson asks: Have these practices and recommendations anything
in common and do they incorporate any of the options and positions
that were put forward by our contributors? He observes that what is
uniformly impressive in this IOM Report is its bleak pragmatism and
scant reference to political and economic ideologies and to social strategies. Architects and implementers of public policy toward the elderly,
and their professional advisors, appear preoccupied with process, with
intricate devices – commonly monetary and regulative – for coping with
overwhelming conditions. There is little historical retrospect on cross-
national comparisons: Was it, in hindsight, the best advice that the
Institute of Medicine made to the nation, in 1977, when it concluded
that there was no need of a specialty in geriatric practice, given the
persistence of inadequacies in nursing home care? The IOM Report

contains no analysis of possible professional and institutional dysfunctions that makes the social and medical consequences of our demography worse than it would otherwise be. The Report advocates systems perseveration in the current mode. The only innovation in Medicare funding, despite its budgetary strictures otherwise, has been the extension of benefits to support cardiac transplantation, surely an expression of today's dominant interests of high technology in medicine.

It is plain that, as with our industrial perspective, we are dangerously limiting ourselves to merely adapting to short-term exigencies. It appears to be a tacit fact of our present political as well as managerial culture, in which external events like demography and foreign competition are pled and proffered as "reasons" for not doing better or differently.

Faced with the same undeniable irreversibility of the demographic momentum of the population with all of its consequences, the contributors to this volume were concerned with the long-range view, with weighing future costs and choices; they almost invariably assume *changeability* in the way we do things, conduct ourselves, reward ourselves, and organize our activities. We may not have to proceed too long, or wait much longer, for their reflections and contemplations to appear the more practical and certainly the more humane.

It is apparent from this Introduction that the symposium which led to this volume brought together various disciplines to assess and discuss our individual, community, and societal responsibilities as average life expectancy and the proportion of elderly increase. Many contributors have addressed various conceptual traps or possible misconceptions of aging, dependency, rationing and cost containment, medical heroics, the so-called "burden" of the aged, and the meaning of competency as we age.

The volume also outlines policy choices and organizational models to meet current and future challenges as we attempt to create a more humane society for all of us. Although the policy debates revolve around both collective and individual responsibilities, distributive justice, and social class inequities, it is immediately apparent that these important issues do not capture entirely the organizational, political, and economic complexities which will face all societies in the 21st century.

The U.S. is presently focussed on cost containment and the goal of economic efficiency. Consequently, in part, both quality of care and life have shifted slightly into the background. In addition, various leaders in medicine, the social sciences, and the humanities have realized that the traditional biomedical model, albeit very powerful, does require reform if the delivery of health care is to be accessible to children and adults of all ages. Pediatric practice, and more recently geriatric medical practice, although not without contradictory tendencies, represent a social thrust, that is, an attempt to modify the biomedical model. Witness the recent broad focus of WHO's new initiative: Health for All by the Year 2000!

Does today's attention to prospective payment schemes as well as the rapid growth of capitation payment models mean the U.S. is moving toward a true *health* care system in contrast to a *medical* care system? Because significant public funding sustains the present profit motive within the medical care arena, there are strong vested interests at work which tend to preserve the status quo. It is hard not to pay people well for the tasks they prefer to do, and shift funds to the presently unprofitable things people ought to do. The economics, like the systems we devise, actually *reflect* rather than significantly *determine* our moral choice.

The care of the elderly – like all U.S. health care – is thus in a state of transition in terms of its economic base and the moral choices that underlie the economic system. Everyone involved in this transition phase – patients, physicians, hospital managers, employers, third-party payers, and state and federal bureaucrats – will surely affect the final outcome, however varied it is in design when seen in retrospect from the 21st century. Broadly conceived, this volume focusses on the often tacitly value-laden dimensions of the care of the elderly. The contributors have adroitly posed, if not fully resolved, the salient issues which will certainly become the basis for the value conflicts of the 21st century.

December, 1986   RICHARD A. LUSKY

and The Editors

BIBLIOGRAPHY

1. Bureau of the Census: Projections of the Population of the United States, 1983–2080. *Current Population Reports*: 1984, Series, P–25, No. 952. Public Health Service, Hyattsville, Maryland, pp. 38–106.

2. Fries, J. F.: 1983, 'The Compression of Morbidity', *Milbank Memorial Fund Quarterly/Health and Society* **61**, 397–413.
3. Institute of Medicine (Committee on Nursing Home Regulation): 1986, *Improving the Quality of Care in Nursing Homes*, National Academy Press, Washington, D.C.
4. National Center for Health Statistics. *Health United States*: 1982, DHHS Pub. No. (PHS) 83–1232. Public Health Service, U.S. Government Printing Office, Washington, D.C.
5. National Center for Health Statistics. M. G. Kovar: May 9, 1986, 'Aging in the Eighties, Age 65 Years and Over and Living Alone, Contacts with Family, Friends, and Neighbors', *Advance Data from Vital and Health Statistics*, No. 116. DHHS Pub. No. (PHS) 86–1250, Public Health Service, Hyattsville, Maryland.
6. Ouslander, J. G. and Beck, J. C.: 1982, 'Defining the Health Problems of the Elderly', *Annual Review of Public Health* **3**, 71–76.
7. Rice, D. P. and Feldman, J. J.: 1983, 'Living Longer in the United States: Demographic Changes and Health Needs of the Elderly', *Milbank Memorial Fund Quarterly/Health and Society* **61**, 382–383.
8. Sheppard, H. L. and Rix, S. E.: 1979, *The Graying of Working America*, Free Press, New York.
9. U.S. Department of Health and Human Services: 1986, *Aging America Trends and Projections*, (1985–86 Edition), U.S. Government Printing Office, Washington, D.C.
10. Vaupel, J. W. and Gowan, A. E.: 1986, 'Passage to Methuselah: Some Demographic Consequences of Continued Progress Against Mortality', *American Journal of Public Health* **6**(4), 430–433.

EDITORS' ACKNOWLEDGEMENT

On the occasion of the publication of volume 25 of the *Philosophy and Medicine* series, the Editors wish to extend their gratitude to those persons who gave their energies and talents to this on-going research enterprise. In addition to the extensive list of contributors to the Series, many colleagues provided very generous assistance, many doing so without compensation except for the knowledge and satisfaction that they were participating in a scholarly endeavor whose primary purpose was and is to establish a new discipline – the philosophy of medicine – on a firm foundation, and to sustain future scholarship and enhance medical practice by making significant research materials available to future generations of readers and scholars, especially faculty in medicine and philosophy.

In addition to the authors, editors, research assistants, and office staffs – at the Center for Medicine, Ethics and Public Issues of The Baylor College of Medicine in Houston, Texas, and the Department of Community Medicine and Health Care of the School of Medicine of the University of Connecticut Health Center in Farmington – we personally take this opportunity to extend our appreciation to Mr. Jan Hattink, Mr. Martin Scrivener, and the staff and publishers of D. Reidel Publishing Company (a member of Kluwer Academic Publishers) in Dordrecht, Netherlands, and the typesetter in Manila, Philippines.

Finally, we wish to thank our parents, without whose inspiration our careers would never have taken shape, and our wives, Susan E. M. Engelhardt and Lorraine Langlois, for sharing so enthusiastically in this long-term endeavor.

June 3, 1987   TRISTRAM ENGELHARDT, JR.
                   STUART F. SPICKER

SECTION I

UNDERSTANDING THE BIOLOGY AND EPIDEMIOLOGY OF AGING

JACOB A. BRODY

THE BEST OF TIMES/THE WORST OF TIMES:
AGING AND DEPENDENCY IN THE 21st CENTURY

Can we resist the metaphor of the "wonderful one-hoss-shay" alluded to by Fries in his article 'Aging, Natural Death and the Compression of Morbidity' [13]? We all yearn to live in perfect physical and mental harmony with ourselves and our surroundings until some long-postponed event suddenly snatches away our lives. How stark in contrast are W. B. Yeats' lines in his poem 'Sailing to Byzantium':

> Consume my heart away; sick with desire
> And fastened to a dying animal
> It knows not what it is.

My commentary suggests that the human body increasingly outlives its parts – its knees, its back, its eyes, its ears, and its mind.

Life span, which is best defined as the theoretical survival potential of a particular species, can be no less than 110 to 115 years for man [28]. This assumption is based on empirical evidence that some humans have survived that long, rather than on theoretical projections of the ability of human protoplasm to survive. In view of man's predilection for scientific busyness it is likely that life span will exceed the current estimate, perhaps by a considerable amount. Life expectancy is the average observed years of life from birth or any stated age. It is not a theoretical value but is based on accumulated data. In the United States life expectancy is now approximately 71 years for males and 78.3 years for females, an average of about 74 years for both sexes according to *Current Population Reports* [32]. The legitimate concern is clearly the discrepancy between life expectancy and life span – about 40 years, more than half the current life expectancy.

MORTALITY REDUCTION – WHAT CAUSED IT, AND WHAT IS
PERPETUATING IT?

Since 1900, Americans have eschewed dying. In 1900, only a quarter of the population survived to age 65 [5]. At present almost 70 percent

3

Stuart F. Spicker, Stanley R. Ingman, and Ian R. Lawson (eds.),
Ethical Dimensions of Geriatric Care, 3–21.
© *1987 by D. Reidel Publishing Company.*

achieve this age, and the figure for those 80 and over is 30 percent and increasing. Within a very few years, about half the population will live more than 80 years. The pattern of reduction in mortality has not been smooth over time or easily explainable. In Figure 1 (top), we note that the mortality rate per 100 000 for all ages fell from 1720 to 870 between 1900 and 1979. Half the remarkable decrease observed this century was complete by about 1920. This suggests that the hand of the physician or the social planner was not prominent in the period during which the greatest mortality reduction occurred. Instead, it appears that improved living environments and better sanitation were probably the major factors in the decline.

Among those 65 and over, as Figure 1 (bottom) shows, mortality had declined barely 5 percent by 1920, an indication that the older population was resistant to the factors that had caused the conservation of life in the general population. Those 65 and over, however, caught up by about 1945, when one-half the decline in mortality for their age group in this century had been completed. The era between 1920 and 1945 was an eventful one in American history, including such notable events as World War II, the Great Depression, the beginning of Social Security, and Prohibition. While one may speculate that any number of events caused the decline in mortality by 1945, those related to specific medical or social interventions are not likely to have been major contributors.

During the period from 1945 to about 1968, when medical and social factors should have been effectively reducing mortality rates for the entire population, the overall rate was generally unchanging with an actual increase in mortality for those aged 65 and over (Figure 1). This effect reassured policymakers and planners and, in a sense, lulled them into a false assumption of stable predictability (*The Change in Mortality Trends in the U.S.* [24]).

Since about 1968, as Figure 1 shows, mortality rates have followed a persistent decline with only minor perturbations. It is clear from the figure that overall mortality has not been declining at the same rate as mortality among those 65 and over. We are observing an amazing decline which is being enjoyed more by those in the older age groups, and whose result has been a disproportionate accumulation of the oldest and most vulnerable sector of our population.

Can this recent trend be ascribed to medical and social interventions? There is no doubt that some of the things we are doing are having an effect. Perhaps the most powerful medical thrusts have occurred in the

Figure 1

Mortality Rates for Years 1900—1979, by Sex (All Ages)

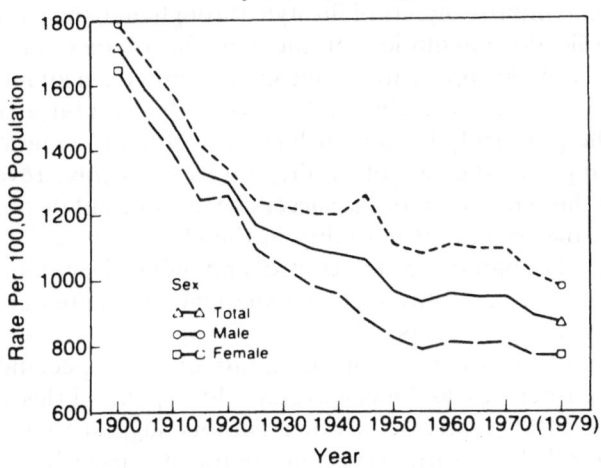

Mortality Rates for Years 1900—1979, by Sex (Ages 65 and Over)

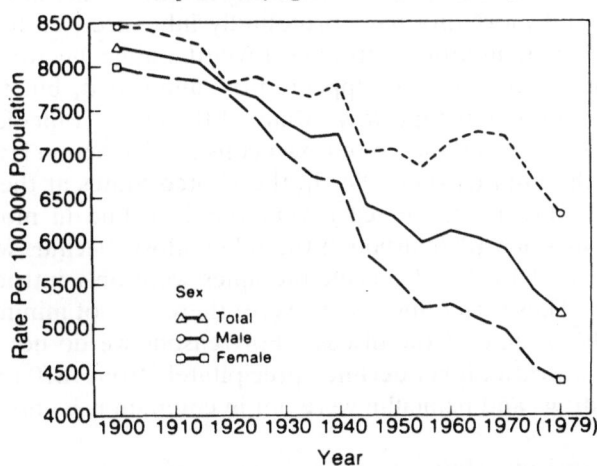

Source. National Center for Health Statistics (U S Public Health Service), various annual volumes of "Vital Statistics of the United States"

reduction of blood pressure and the improvement of medical care through Medicare and other health-insurance programs. Sociomedical and public-health efforts have included a reduction in smoking and the promotion of improvements of lifestyle through diet and exercise. While the scientific documentation of the benefits of diet and exercise is fragile, we would suppose that some good comes from an intelligent and moderate approach to eating and activity. But careful analysis of the data on the powerful effectors of human health and longevity, such as not smoking and the use of antihypertensive agents, reveals that in neither of these parameters has success been sufficient or timely enough to explain the decline of mortality in the elderly since 1968 [30]. The major impact of blood pressure control and reduced smoking is yet to be fully realized. We are thus likely to experience even greater gains in life expectancy in future years.

In his arguments concerning the limits of life expectancy which he sometimes refers to as "ideal average life span," Fries arrives at a critical age of 85 years. Most other authors suggest that this is a low estimate ([22], [4], [26]). Actuarial approaches used by Fries and, of course, in the projections of the Social Security Administration serve as the cornerstone of major policy decisions. My own discomfort with this approach is reflected in the title of this section, which raises questions for which I do not believe we have satisfactory answers.

Major shifts that have occurred in the last 150 years in the causes of mortality and morbidity have profoundly influenced both their occurrence and contemporary attitudes toward public health and policy. Important diseases have disappeared or diminished, but for the most part we do not understand why. Many of the greatest medical triumphs are not easily attributable to our own constructive action [10]. Tuberculosis was the number-one killer in the United States at the turn of the century. In Figure 2, we see a very rapid decline in mortality from tuberculosis since 1856 although the tuberculosis bacillus was not even identified until 1882 and specific therapies were only introduced in the late 1940s. These scientific events were, therefore, of minimum importance in the decline of the disease. For reasons we do not understand, deaths from scarlet fever declined precipitately from 1870 to 1900, while the sulfa drugs and penicillin were not in common use until after World War II.

The surprising shifts in diseases are not confined to diseases of infectious etiology. The rate of mortality from accidents among the

Figure 2
RESPIRATORY TUBERCULOSIS IN THE
TOTAL POPULATION

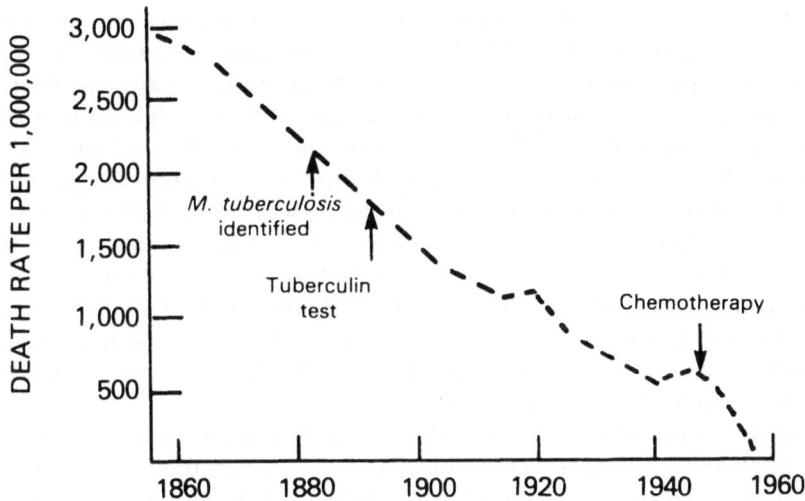

Source: Mean annual death rates from various diseases, England and Wales, 1860-1960. From Winikoff B: *Nutrition and National Policy*, Cambridge, MIT Press, 1978, pp. 444-445.

elderly has been declining rapidly since 1950 (*Handbook of the Biology of Aging* [6]). For reasons poorly understood, conversion hysteria, one of Freud's most frequent diagnoses, seems to have vanished. Mortality from stroke and stomach cancer has been declining since at least 1950 in the United States and perhaps for a much longer time [3].

A phenomenon that has had a great impact on health trends and current attitudes is the remarkable decline since the late 1960s in mortality from ischemic heart disease [25]. In my opinion, it has created a false sense of optimism about the extent to which we are in control of our survival ([30], [33]).

I would argue that heart attacks are somehow a 20th-century medical chimera. There is no real mention of the classical heart attack or myocardial infarction before 1900: the generally accepted first reference is Herrick's in 1912 [17]. We could easily claim that once again the doddering medical profession simply missed the boat and has for centuries been calling heart attacks indigestion or heavy catarrh.

The extraordinarily dramatic nature of a heart attack, with its crushing chest pain and pain radiating down the left arm makes it indeed a notable event. Even if physicians missed the diagnosis, surely our authors and playwrights would have made numerous references to an event so frequent and dramatic as we now witness. There are, however, no good descriptions of heart attacks in a literature that has no lack of superb accounts of gout, epilepsy, apoplexy, and tuberculosis. A plausible scenario is that acute myocardial infarction was created by circumstances arising early in this century, probably associated with the introduction of cigarette smoking and different types and amounts of fat in the diet. The combination was not as lethal for females as for males.

Mortality from heart attacks is now declining in the United States. This decline is not being observed throughout the world and notably in countries such as Sweden [35] and Japan [20] where life expectancy is greater than in the United States.

Failure to recognize that something very important is missing from our knowledge about the cause of heart attacks and their appearance and decline within this century has produced policies and thinking that are risky and incompatible with facts. A more helpful explanation of the heart-attack phenomenon would point out that the disease was introduced into the population during a time when most causes of death were declining. As we eliminated acute and infectious diseases early in the century, the dominant killer became heart disease and it still is. Half of all mortality is caused by heart disease of which a large fraction is the result of acute myocardial infarction. As Figure 1 shows, overall mortality declined steadily until after World War II. Then there was a plateau and even a rise in overall mortality, which may have been the result of a slowing-down of the primary forces that had reduced mortality in the early part of the century, and an accompanying increase in the mortality from heart disease. Acute myocardial infarction, however, appears to have been a chance phenomenon which, while exerting serious damage, was dependent for its maintenance and increase on complex factors, certainly including smoking and diet. The factors, whatever they are, affect men more than women, and their relation to the causes of disease in humans seems tenuous. Thus, the rapid decline in deaths from acute myocardial infarction commenced in the late 1960s at a time when only modest changes in smoking patterns and diet had occurred. It is certainly possible that acute myocardial infarcts will markedly decline or vanish, as have many other serious human maladies such as those mentioned above, because they can be preserved only under unusual circumstances.

We are faced with the fact that older people are living longer [5]. It is likely that this phenomenon will continue, at least in part because the major groupings of diseases, both infectious and chronic, are being eliminated or postponed. I suggest that the age-specific mortality rates in the United States since 1900 would have shown a general, persistent, and much smoother decline over the entire 20th century had it not been for the remarkable appearance of mortality from acute myocardial infarction.

The driving force in the extension of life expectancy appears more closely to fit a pattern of the release of something intrinsic in humans rather than a decline in cause-specific mortality. For at least the past century we have been moving nearer to our genetic potential. There is an analogy, and almost surely a relationship, between increasing survival and the consistent observation that for more than a century there was a steady increase in average height of successive generations. The factors likely to have led to the extension of growth are the elimination of frequent and severe infectious diseases and better nutrition in early and formative years. There is evidence, in some populations, that growth may have reached a plateau, either temporarily or, perhaps, permanently [1].

Biologically, I think we can at least postulate that we are producing a person better equipped for survival, since with each successive generation the cohort children more closely approach the maximum human growth potential. This phenomenon has set up cohorts with successively improved trajectories for survival. By studying the past history of disease and mortality data we can make only fragile guesses about human survival, as Fries does with his actuarial interpolations. No other species has ever had the potential for controlling its environment, and so we simply do not know the limits of life expectancy. There are obviously no empirical data of great value to instruct us on the limits of this unprecedented happening.

DISABILITIES

Life satisfaction and quality of life are critical areas in which scientific documentation is elusive and, at times, inaccessible. The most useful surrogate measurement is perhaps disabilities, which frequently arise from illness or morbidity and are also difficult to measure. In the sections above we have demonstrated that life expectancy is increasing at a remarkable rate that is higher among the elderly than in other age groups. Thus, as people live longer we are faced with the problem of making their longer lives more satisfactory. Our key tool is to postpone, eliminate, or mitigate the burden of disability.

Most recent commentators have not looked at the future with great enthusiasm. Schneider and Brody stated, "If current demographic trends continue we will clearly be faced with increased numbers of people at advanced ages. The unknown variable will be the health of this group. If the health of this group in the future is not considerably different from the health of the present cohort, a huge proportion of the population will be suffering from chronic diseases" [28]. Feinleib said, "We must bear in mind that as we keep people alive longer, we will be performing many more procedures for diseases associated with increasing age – more prostate operations, more cataract operations, more hip replacements, etc. We will be obliged to devote a greater proportion of our resources to restorative procedures for the elderly" [12]. Jacob S. Siegel, retired senior statistician for demographic research and analysis, Population Division, U.S. Bureau of the Census, and perhaps the dean of the demographers of the elderly, sees the future very bleakly indeed: "This development suggests the desirability of pursuing intensive research into the complex bioethical issues relating to right to die and suggests the need to explore the concept of life with dignity and the right to die as basic human rights" [29]. A gloomy perspective of the health status and quality of life of the elderly is offered in Gruenberg's 'The Failure of Success' [16] and Kramer's 'The Rising Pandemic of Mental Disorders and Associated Chronic Diseases and Disabilities' [21]. Both address the present and future conditions of the elderly, observing that since chronic diseases increase with age, postponement of death through medical intervention in a growing population of the elderly will inevitably increase both the prevalence and the absolute number of sick old people.

Fries, in a more optimistic vein, makes a series of questionable assumptions based on his assertion that human life expectancy will not exceed 85 years. He states, "This article discusses a set of predictions that contradict the conventional anticipation of an ever older, ever more feeble, and ever more expensive-to-care-for populace. These predictions suggest that the number of very old persons will not increase, that the average period of diminished physical vigor will decrease, that chronic disease will occupy a smaller proportion of the typical life span, and that the need for medical care in later life will decrease" [13].

Fries' position is difficult to analyze, since as we have pointed out life expectancy in the United States is still 10 to 12 years short of his 85-year average. Even Fries in his later writing acknowledges that we lack empirical data on the physical vigor when half of the entire population will survive to age 85 and the rapidity with which they will die – either following the waning of physical vigor or at age 85. "Conclusive data are difficult to come by because data on the incidence of markers of morbidity have not been systematically collected, nor do we have even in cross-sectional studies, prevalence figures for the 'quality of life.' Indeed, it is not even clear what measures of morbidity should be used" [14].

Further, although only one percent of our population is now over age 85, recent data indicate that this group is experiencing the greatest increase in life expectancy ([28], [32]). Manton [22] and Myers and Manton [23] show that while the average age at death has been increasing rapidly since 1962, the variance in the older fraction has been stationary or increasing slightly rather than decreasing as would be necessary to conform to Fries' notion. The survival curve beyond age 85 is becoming less rectangular and more horizontal. As we have pointed out above, we certainly have no means at hand to determine when this trend toward greater longevity will end.

If people are living longer, logic suggests that they must have achieved older age because they have gotten healthier. Some data on improved health status among successive aged cohorts are accumulating in studies being conducted by Svanborg and his group in Goteborg, Sweden; by Branch and his coworkers in Massachusetts; and by Reed and Benfante in Honolulu working with Brody and his group at the National Institute on Aging. These and perhaps other studies that suggest a pattern of increasing years of good health are promising, but the data are not overwhelming.

Katz et al. [19] are developing a promising approach to the measurement of health in a population in terms other than death. They advance the concept of "active life expectancy," the endpoint of which is loss of independence in the activities of daily living. Their data, based on small numbers, show a very substantial decline in active life expectancy between age 65 and 84. Also relevant is the finding that men died shortly after loss of independence, while women lived on with diminishing functional abilities. The suggestion is that we are not, as Fries states, compressing morbidity in any effective way other than in the unacceptable premature mortality of males. Should male patterns more closely resemble those of females, the data indicate that the number of years lived after active life expectancy would increase – surely the opposite of Fries' hypothesis. Theoretical papers by Schatzkin [27] and Manton [22] convey similar implications.

Impediments to the documentation of health status over time are numerous. Colvez and Blanchet [9] found a disturbing rise in recent years in morbidity and disability in persons of all age groups, but particularly in middle life. Their data were based on the National Health Interview Survey (NHIS) which has been conducted annually since 1957 by the National Center for Health Statistics (NCHS). Similar results were shown in a large recent Veterans Administration study conducted by Louis Harris and Associates, Inc. [34]. Wilson [36] of the NCHS, in an editorial comment on the Colvez and Blanchet report, questions the ability "of health statistics to reflect what might really be happening to health status." He offers a series of possible explanations for the observed increases in morbidity and disability. These range from artificially high results, caused by an increased sensitivity of the screening instrument to a change in attitude and self-reporting as a result of more liberal retirement benefits for health reasons, to a real increase due to improved survival of the chronically ill. This accurate but untidy array of options concerning the measurement of health status emphasizes the fact that we do not have sufficient data to reach any conclusion about whether people are indeed enjoying better health.

Although we have excellent survey tools to determine health and functional status and to measure degree of disability, changes in the way people respond over time limit our ability to determine trends. The level of knowledge, awareness, and sophistication in matters of health has been increasing rapidly. For instance, as education about hypertension has increased, more people have had their blood pressure determined. Since hypertension is a key item in a health-status questionnaire, the rate of reported hypertension has of course increased ([9], [34]).

Longitudinal studies in which subjects are routinely examined over time avoid some of the obvious weaknesses of the sequential survey approach. Longitudinal surveys, however, are few in number and because of the cost and logistical problems can follow only relatively small cohorts.

The limitations of surveys and longitudinal studies are apparent. But this is the study of mankind. We must be able to monitor morbidity, disability, and even the quality of life.

DISCUSSION

A practical reality dominates our thinking. Because of the combination of increasing life expectancy and the maturation of the "baby-boom" population, extraordinary shifts toward an increased number of older Americans are inevitable. Within about 40 years (by 2025), the population 65 and over will have risen from the current 11.3 to 20 percent and will comprise about 60 million individuals [32]. By the year 2050, the population 85 and over will have risen from about 2 million to more than 16 million [32]. Sometime around 1990, more than half of the people age 65 and over will be age 75 and over. When we lament that our medical and social systems cannot provide for those 65 and over at present, we are taking a very short view. It is after age 75 that health problems increase sharply (*The Teaching Nursing Home: A New Approach to Geriatric Research, Education, and Clinical Care* [8]). I am concerned about the effect that Fries' remarks will have on health planners by suggesting that the number of very old persons will not increase, that diminished physical vigor will decrease, and these factors will result in a diminished need for medical care in later life [13].

A concerned government should focus on the realities of the onrushing future. There seems to be evidence that life expectancy will increase at least to age 85, but there appear no obvious barriers to an even more prolonged survival. In some societies (Japan/Sweden) women are already approaching a life expectancy of 85 years in the presence of very high levels of hypertension among the elderly ([31], in press). What will be the effect when the already available antihypertensive agents are widely used among these women? I would presume 5 to 10 years of increased survival.

I will use three specific examples to illustrate physical, mental, and social realities that will occur unless major, unforeseen advances in science and health management develop. In 1980 there were approximately 200 000 people with hip fractures whose median age was 79. By the year 2000 the number will rise to 330 600. Figure 3, which uses census projections, illustrates that by 2050 this number will have risen to more than 650 000, including 225 000 age 75 to 84, and 340 000 over age 85.

In 1980 there were more than 2 000 000 people with Alzheimer's disease or related disorders (Figure 4) whose mean age was about 80 years. By the year 2000 the number will rise to 3 800 000. By 2050 there will be more than 8 500 000 of whom almost 5 000 000 will be over age 85 and over 2 700 000 between 75 and 84.

There are now about 1.1 million people age 65 and over in nursing homes whose median age is about 80. By the year 2000 the number will rise to 2 200 000. In 2050 (Figure 5), a conservative estimate projects a total of 5 400 000.

As a society, we must address the implications of our rapidly aging elderly population. I do not find useful Fries' suggestion that chronic disease is simply part of the road to "emergence of a pattern of natural death" [13]. There is no evidence that natural death exists. The evidence overwhelmingly indicates that people are surviving to older ages when chronic diseases or acute disease can produce death at a lower physiologic threshold. These diseases also can produce disability without death, a state we increasingly observe.

Of course, as Fries urges there is considerable value in health promotion for disease prevention and in the value of making people more involved in self-care and in achieving wholesome lifestyles. We may, however, face a credibility gap in furthering these goals by overstating our knowledge in areas in which scientific evidence is weak or lacking ([5], [33]).

In the face of the burgeoning number of old and very old people we must admit to ourselves that scientific knowledge of the cause and prevention of most chronic diseases is limited [4]. Only a fraction of the cancers, heart diseases, strokes, neurologic diseases, and arthritides are understood, and we know virtually nothing about the causes and prevention of the dementias of old age. No one doubts that smoking and excessive drinking are harmful. Data on lowering blood pressure are robust, and at least one study has shown a preventive effect on lowering cholesterol level [18]. Data are softer relating to exercise, weight control, special dieting, and salt restriction in the absence of hypertension. Nor do we know how great would be the reduction of morbidity and disability if we produced a population of paragons ([5], [2]). If we are to

Figure 3

PROJECTED NUMBER OF HIP FRACTURES
ANNUALLY IN THE U.S. BY AGE: 1980-2050

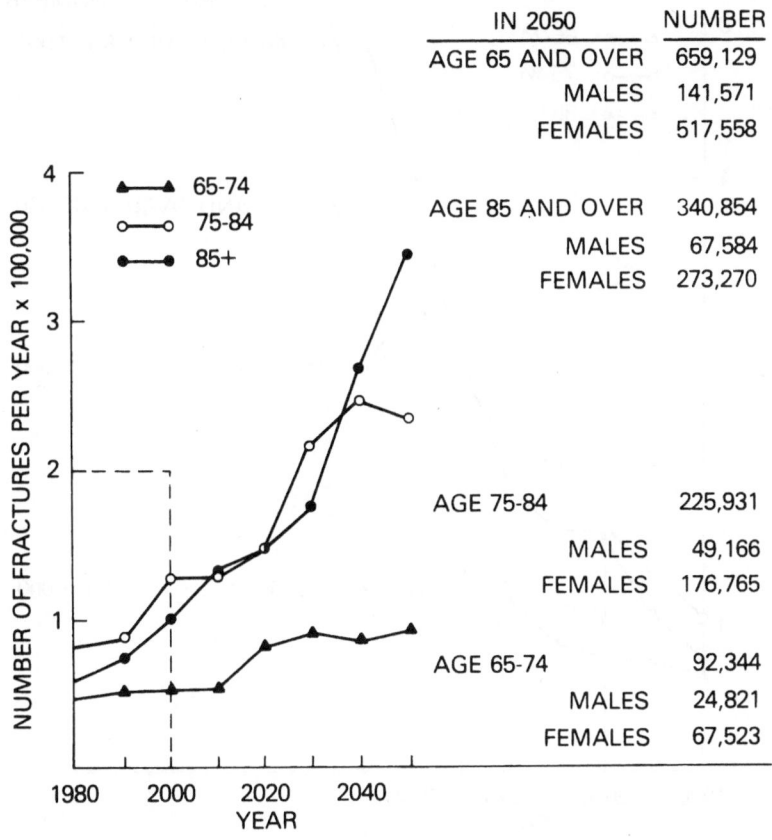

IN 2050	NUMBER
AGE 65 AND OVER	659,129
MALES	141,571
FEMALES	517,558
AGE 85 AND OVER	340,854
MALES	67,584
FEMALES	273,270
AGE 75-84	225,931
MALES	49,166
FEMALES	176,765
AGE 65-74	92,344
MALES	24,821
FEMALES	67,523

Source: NCHS and U.S. Bureau of Census projections

maintain our credibility, we have an obligation to admit our ignorance of causation and our tendency to create guilt and anxiety among the elderly in areas in which the benefits of specific preventive techniques have not been scientifically documented [33].

Figure 4

PROJECTED NUMBER OF DEMENTED PERSONS IN THE U.S. BY AGE: 1980-2050

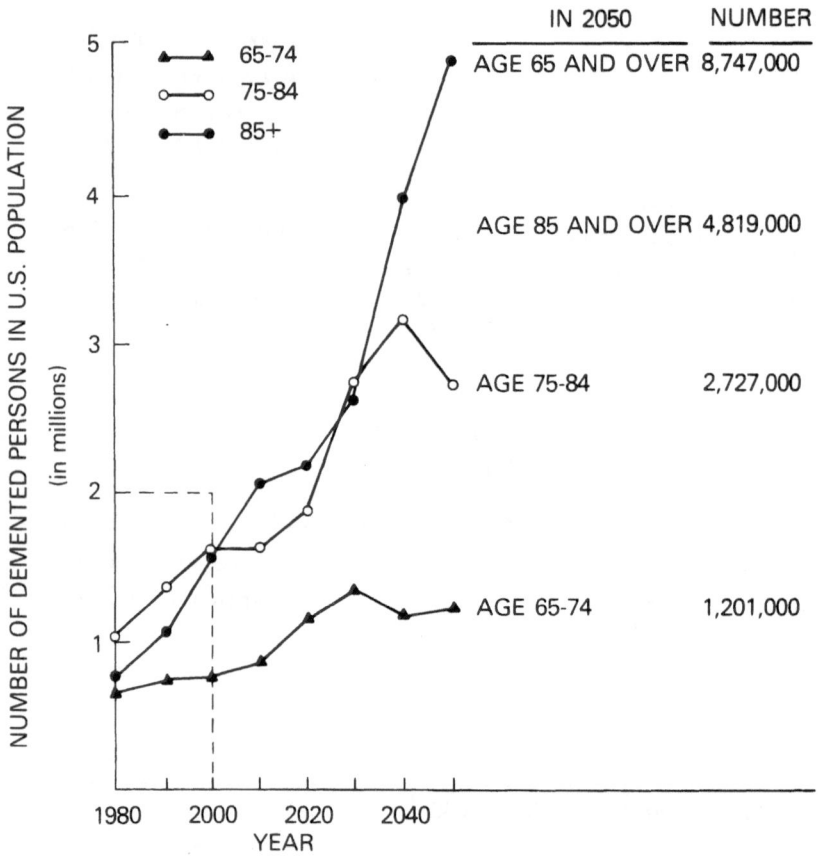

Source: NIA prevalence estimates and U.S. Bureau of Census projections

CONCLUSION

I think we must accept the reality that the human body outlives its joints, its cognitive and memory capacities, its vision, its hearing, and the numerous other causes of disability. Fries' concept of homeostasis applies almost entirely to vital organs and mortality, but not to morbid-

Figure 5

PROJECTED NUMBER OF NURSING HOME RESIDENTS IN THE U.S. BY AGE: 1980-2050

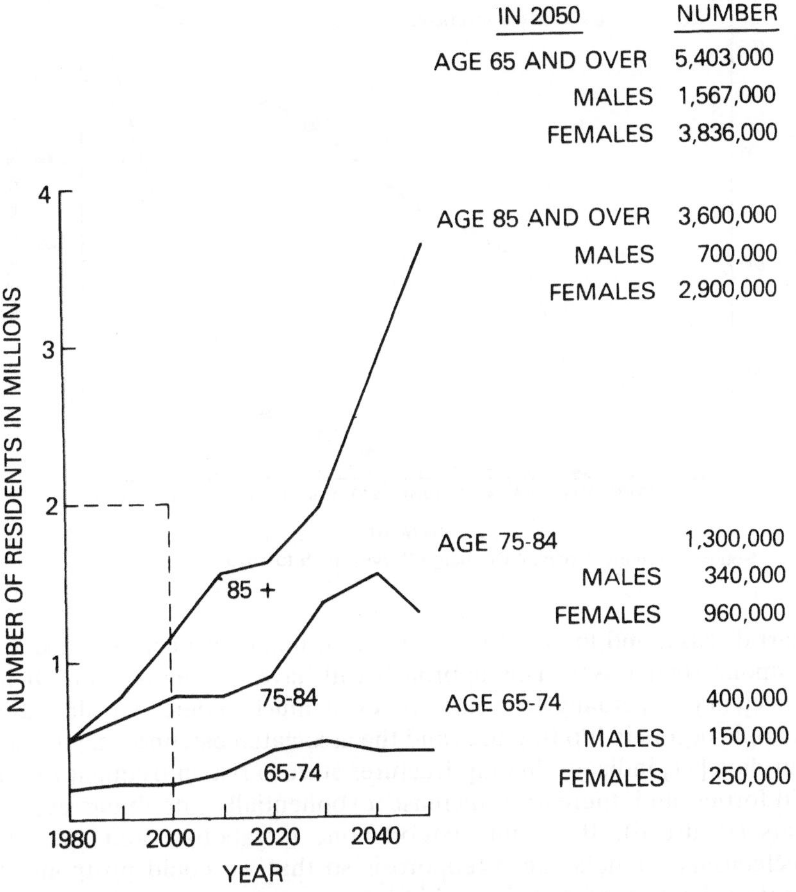
IN 2050
NUMBER
AGE 65 AND OVER 5,403,000
MALES 1,567,000
FEMALES 3,836,000
AGE 85 AND OVER 3,600,000
MALES 700,000
FEMALES 2,900,000
AGE 75-84 1,300,000
MALES 340,000
FEMALES 960,000
AGE 65-74 400,000
MALES 150,000
FEMALES 250,000
NUMBER OF RESIDENTS IN MILLIONS
4
3
2
1
85 +
75-84
65-74
1980
2000
2020
2040
YEAR

Source: NCHS and U.S. Bureau of Census projections

ity, disability, or the quality of life [14]. With aging there is little likelihood that morbidity will be compressed. Our strategy must be to recognize those conditions which become more prevalent with aging – such as blindness, deafness, Alzheimer's disease, acute and chronic

Figure 6

AGE-SPECIFIC HIP FRACTURE INCIDENCE AMONG WHITE WOMEN

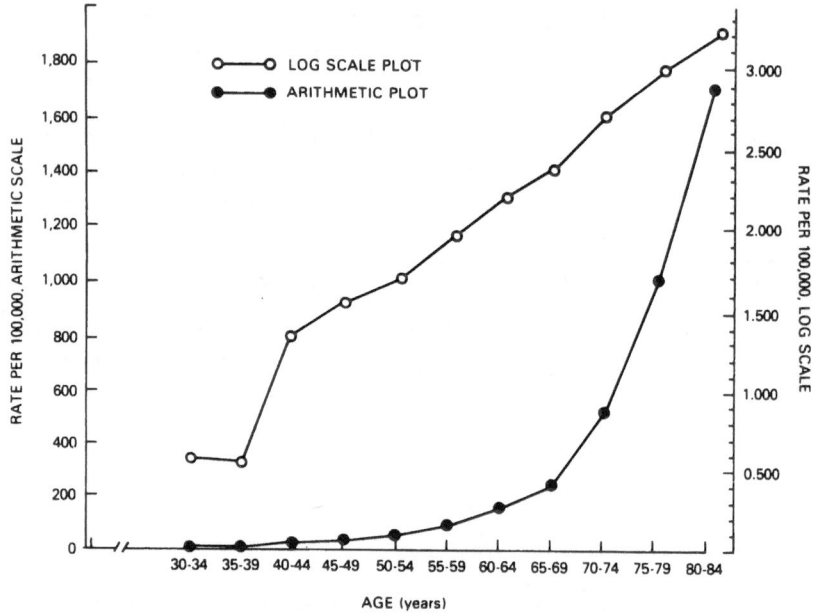

Source: National Hospital Discharge Survey 1975 to 1979.

heart disease, and hip fracture. The major impact we can produce is to postpone their onset. This approach will have enormous value if life expectancy is actually 85 years, or even much older. An illustrative example occurs in hip fractures and the associated osteoporotic process. Our data [11] indicate that hip fractures start to rise in frequency in the mid-forties and thereafter increase exponentially, doubling every 6 years (Figure 6). If we understood bone metabolism and knew the mechanisms for delaying osteoporosis so that we could postpone the onset of the exponential rise in hip fracture occurrence for 6 years, we would in fact reduce the incidence of hip fracture by 50 percent [7]. Research directed at the precursors of disability is the most effective tool we have to prevent chronic disease and attendant disabilities.

Research directed at developing and exploiting the emerging data on elderly populations must receive the highest priority. As I have emphasized, the increase in life expectancy we are witnessing is unprecedented in human history. The implication of the systematic alteration between 1900, when 25 percent of the population survived their 65th year, and 1980, when almost 70 percent survived their 65th year, can be understood only through intensive analysis of the emerging new data. We have not yet built up an adequate body of knowledge of human natural history beyond age 65. In our state of ignorance, making assumptions with major policy implications is precarious. When we speculate, in good faith but based on projections that human history has already bypassed, we run the risk of ignoring Hippocrates' most basic admonition: "The physician must be able to tell the antecedents, know the present, and foretell the future – must meditate these things, and have two special objects in view with regard to diseases, namely, to do good or to do no harm" [15].

University of Illinois
Chicago, Illinois

BIBLIOGRAPHY

1. Abraham, S., Johnson, C. L., and Najjar, M. F.: 1977, 'Weight by height and age of adults 18–74 years: United States, 1971–74', *Vital and Health Statistics*, No. 14, U.S. Government Printing Office, Washington, D.C.: pp. 1–12.
2. Benfante, R., Reed, D., and Brody, J. A.: 1985, 'Biological and social predictors of health in an aging cohort', *Journal of Chronic Diseases* **38**, 385–395.
3. Brock, D. B. and Brody, J. A.: 1984, 'Statistical and epidemiologic characteristics of the United States elderly population', in R. Andres, E. L. Bierman, and W. R. Hazzard (eds.), *Principals of Geriatric Medicine*, McGraw-Hill, New York, pp. 51–71.
4. Brody, J. A.: 1982, 'Life expectancy and the health of older people', *Journal of the American Geriatrics Society* **30**, 681–683.
5. Brody, J. A.: 1984, 'Facts, projections, and gaps concerning data on aging', *Public Health Reports* **99**, 468–475.
6. Brody, J. A. and Brock, D. B.: 1984, 'Epidemiologic and statistical characteristics of the United States elderly population', in C. E. Finch and E. L. Schneider (eds.), *Handbook of the Biology of Aging*, 2nd ed., Van Nostrand Reinhold, New York.
7. Brody, J. A., Farmer, M. E., and White, L. R.: 1984, 'Absence of menopausal effect on hip fracture occurrence in white females', *American Journal of Public Health* **74**, 1397–1398.
8. Brody, J. A. and Foley, D. J.: 'Epidemiologic considerations', in N. List, M. Ory, E. L. Schneider, C. Wendland, and A. Zimmer (eds.), *The Teaching Nursing Home: A New Approach to Geriatric Research, Education, and Clinical Care*, Raven Press, New York, in press.

9. Colvez, A. and Blanchet, M.: 1981, 'Disability trends in the United States population 1966–76: Analysis of reported causes', *American Journal of Public Health* **71**, 464–471.
10. Ewbank, D. and Wray, J. D.: 1980, 'Population and public health', in Maxcy-Rosenau (ed.), *Public Health and Preventive Medicine*, 11th ed., Appleton-Century-Crofts, New York, pp. 1504–1548.
11. Farmer, M. E., White, L. R., Brody, J. A., and Bailey, K. R.: 1984, 'Race and sex differences in hip fracture incidence', *American Journal of Public Health* **74**, 1374–1380.
12. Feinleib, M.: 1980, 'Recommendations for future research', in S. G. Haynes and M. Feinleib (eds.), *Proceedings of the Second Conference on the Epidemiology of Aging*, NIH Pub. No. 80–969, U.S. Government Printing Office, Washington, D.C., p. 359.
13. Fries, J. F.: 1980, 'Aging, natural death, and the compression of morbidity', *The New England Journal of Medicine* **303**, 130–135.
14. Fries, J. F.: 1983, 'The compression of morbidity', *Milbank Memorial Fund Quarterly/Health and Society* **61**, 397–419.
15. Galen: 1952, 'Hippocratic writings on the natural faculties', in R. M. Hutchins (ed.), *Great Books of the Western World*, Encyclopaedia, Inc., Chicago, p. 46.
16. Gruenberg, E. M.: 1977, 'The failures of success', *Milbank Memorial Fund Quarterly/Health and Society* **55**, 3–24.
17. Herrick, J. B.: 1912, 'Clinical features of sudden obstruction of the coronary arteries', *The Journal of the American Medical Association* **59**, 2015–2020.
18. Hjermann, I., Velve Byre, K., Holme, I., and Leren, P.: 1981, 'Effect of diet and smoking intervention on the incidence of coronary heart disease: Report from the Oslo Study Group of a randomized trial in healthy men', *The Lancet* **2**, 1303–1310.
19. Katz, S., Branch, L. G., Branson, M. H., Papsidero, J. A., Beck, J. C., Greer, D. S.: 1983, 'Active life expectancy', *The New England Journal of Medicine* **309**, 1218–1224.
20. Kimura, N.: 1983, 'Changing patterns of coronary heart disease, stroke, and nutrient intake in Japan', *Preventive Medicine* **12**, 222–227.
21. Kramer, M.: 1980, 'The rising pandemic of mental disorders and associated chronic diseases and disabilities', *Acta Psychiat Scand*, Suppl. 285, **62**, 382–396.
22. Manton, K. G.: 1982, 'Changing concepts of morbidity and mortality in the elderly population', *Milbank Memorial Fund Quarterly/Health and Society* **60**, 183–244.
23. Myers, G. C. and Manton, K. G.: 1984, 'Compression of mortality: myth or reality?' *The Gerontologist* **24**, 346–353.
24. National Center for Health Statistics: 1964, *The Change in Mortality Trends in the U.S.*, Series 3, No. 1, U.S. Government Printing Office, Washington, D.C.
25. Rosenberg, H. M. and Klebba A. J.: 1979, in R. J. Havlik and M. Feinleib (eds.), *Proceedings of the Conference on the Decline in Coronary Heart Disease Mortality*, NIH Pub. No. 79–1610, U.S. Government Printing Office, Washington, D.C., pp. 11–39.
26. Rosenwaike, I., Yaffe, N., and Sagi, P. C.: 1980, 'The recent decline in mortality of the extreme aged: An analysis of statistical data', *American Journal of Public Health* **70**, 1074–1080.
27. Schatzkin, A.: 1980, 'How long can we live? A more optimistic view of potential gains in life expectancy', *American Journal of Public Health* **70**, 1199–1200.

28. Schneider, E. L. and Brody, J. A.: 1983: 'Aging, natural death, and the compression of morbidity: Another view', *The New England Journal of Medicine* **309**, 854–856.

29. Siegel, J. S.: 1980, 'Recent and prospective demographic trends for the elderly population and some implications for health care', in S. G. Haynes and M. Feinleib (eds.), *Proceedings of the Second Conference on the Epidemiology of Aging*, NIH Pub. No. 80–969, U.S. Government Printing Office, Washington, D.C., p. 313.

30. Stallones, R. A.: 1982, 'The rise and fall of ischemic heart disease', *Scientific American* **243**, 53–59.

31. Svanborg, A., Shibata, H., Hatano, S., and Matsuzaki, T.: 'Comparison of Ecology, Ageing and State of Health in Japan and Sweden, the Present and Previous Leaders in Longevity', in press.

32. Taeuber, C. M.: 1983, 'America in transition: An aging society', *Current Population Reports*, Series P-23, No. 128, U.S. Government Printing Office, Washington, D.C., p. 3.

33. Thomas, L.: 1979, *The Medusa and the Snail*, The Viking Press, New York, pp. 165–166.

34. Veterans Administration: 1984, *Survey of Aging Veterans, A Study of the Means, Resources and Expectations of Veterans Aged 55 and Over*, Office of Reports and Statistics, Statistical Policy and Research Service, RSM 70–84–3, U.S. Government Printing Office, Washington, D.C., iv, 52–53.

35. Welin, L., Larsson, B., Svardsudd, K., Wilhelmsen, L., and Tibblin, G.: 1983, 'Why is the incidence of ischaemic heart disease in Sweden increasing?: Study of men born in 1913 and 1923', *The Lancet*, pp. 1087–1089.

36. Wilson, R. W.: 1981, 'Do health indicators indicate health?' *American Journal of Public Health* **71**, 461–463.

TOM BEAUCHAMP

COMMENTARY ON JACOB A. BRODY'S ESSAY

My comment on Jacob Brody's paper is less of a critique than an approach to the moral and policy questions embedded in his paper that are appropriate to this volume. Most are not raised in his paper – or are raised only obliquely. I am also anticipating arguments in some of the later papers in the volume – in particular, the Daniels–Engelhardt encounter.

Some general questions can be raised first about our obligations, if any, to the afflicted elderly. Brody argues that "a huge proportion of the [future] population will be suffering from chronic diseases" if present levels of health care are assumed in studying demographic trends and then are used to project the future. He quotes Feinleib's statement, evidently with approval, that "we will be obliged to devote a greater proportion of our resources to restorative procedures for the elderly." But is this true? If it is, what is the substantive basis of the obligation? Is it an already established obligation, or one that we will have to assume in the future? If, as Brody puts it, "the prevalence and absolute number of sick old people inevitably will rise," does our obligation to them increase or decrease as a result? What would the basis of the claim be in either case?

Brody maintains that "a concerned government should focus upon the realities of the onrushing immediate future" of a projected life expectancy increasing to age 85, or even beyond. But how should this "concern" be manifested, if it should (and no argument is presented that it should)? Why is it a government responsibility, if it is? For example, if we now have 1.3 million people in nursing homes, is alleviation of attached problems a government responsibility? Does the responsibility increase or decrease if there will be 5.4 million residents in the year 2050? Are there alternative policies that would serve us better than direct government intervention and responsibility?

Suppose we have already determined in our society that all human beings are entitled to equal treatment in some respects. Suppose we further agree, as Brody evidently assumes, that they are entitled to at least have basic needs satisfied by society before other members of

23

Stuart F. Spicker, Stanley R. Ingman, and Ian R. Lawson (eds.),
Ethical Dimensions of Geriatric Care, 23–27.
© 1987 by D. Reidel Publishing Company.

society can have a moral right to a higher standard of living. Should we go still further and maintain that all persons in society have a right to have one of their most basic needs – their health care needs – satisfied, before any individual can have a moral right to the life-style now enjoyed by others in the society? Do we also have a moral obligation to pour money into preventive research, which Brody says "must receive the highest priority"?

Some undoubtedly would contend that we do have such an obligation, on grounds that we have an obligation to prevent suffering whenever possible. But what is Brody's view, and what are the implications of his demographic work?

Further, what are the limits of what we should do in the way of providing health-related resources? Is Brody's claim an announcement that the medical age of financial limits has arrived? Is he saying just the reverse? If, as he suggests, 85 years of life expectancy is a low projection, how are we possibly to plan for health care for a population with a retirement age of 65 (even 59 1/2), living in larger numbers to the age of 90 and above? I shall return momentarily to this problem of the impact of early and late retirement schemes.

A major consideration is whether unbearable social costs will follow from the program Brody outlines. How extensive would the financial demands be? Will considerations of justice conflict directly with those of efficiency and utility? Here allocation problems inevitably become central. If one assumes a positive right to health care, as I suppose Brody does, the central question is not whether there is such a right, but the limits to the right. And this problem can only be treated by some principled or procedural way of identifying which or what kinds of health care services we should devise to discharge our obligations.

Anyone who makes decisions about funding health programs and health technologies is confronted with an array of concrete alternatives, each with a nexus of positive and negative factors. A health planner needs an evaluative analytical framework for processing economic data, technology assessments, and public needs. While such an instrument must be compatible with the demands of justice, it must also be capable of handling public policy in a principled rather than an arbitrary way. How would Brody handle these problems for geriatric care and research? Would cost/benefit analysis be widely used as a systematic analytical framework on the basis of which health policy can be developed? Is cost/benefit analysis an appropriate device for the clarification of the overt and covert trade-offs involved in these public policy decisions?

Further, Brody defends the view that "improved living environments and better sanitation were probably the major factors in the decline" of mortality rates. What would this suggest about our policies? Ought health policy to be subordinated to other non-health-related policies? If "medical or social interventions are not likely to have been major contributors" to the decline, as he argues, then what does this conclusion entail about health policy planning for elderly populations? If, as he says, the most powerful impact has been in areas such as anti-smoking campaigns, reduction of tuberculosis and severe infectious disease, the reduction of blood pressure, and better health insurance programs, should we target these or similar areas for special attention? If most of our successful interventions have been public health interventions, as he seems to believe, does this entail that we ought to be placing more of our efforts into public health rather than medical programs – e.g., health education programs or prevention programs?

Brody also contends that our ignorance about the causes and prevention of heart attacks is producing "*policies* and *thinking* which are risky and incompatible with facts"? What facts? Whose policies? How should these facts, whatever they are, be connected to policies? How is that link established, and what ought our policies to be – e.g., policies for the control of heart attacks? Clearly Brody dislikes both the empirical and policy proposals in Fries' work. I assume he rejects the policy implications because he does not believe the empirical claims. But, again, what is the connection between facts and policies in his own account (once Fries' position has been rejected)?

At this point we need to consider Engelhardt's question, which is presented in another essay in the present volume. He suggests that before addressing what would count as a reasonable priority in the allocation of resources, we should ask if one must provide resources for health care? When is it fair and when is it unfair to fail to provide health care? Is it an unjust society if that society makes no provision to help the elderly, but expends its funds elsewhere? Who is responsible, if anyone, for the problems? Suppose our society develops no system for aid. Are we bad Samaritans? Is bad Samaritanism unjust? Does the question turn on whom one thinks owns or is entitled to goods and services? What attitude should be struck toward chronic disease? These are all serious policy problems that it seems.to me Brody's analysis glides over.

Next, we should consider Daniel's problem, also found elsewhere in this volume: Central to justice is a principle of fair opportunity, according to which [in appropriately abbreviated form] no person should be denied social benefits to primary goods on the basis of disadvantaging properties for which the person is not responsible. The justice of social institutions is, in some theories, to be gauged by their tendency to counteract lack of opportunity caused purely by luck of birth (family and class origins), natural endowment and health status, age, and historical circumstances (accidents at some point or over the course of a lifetime). Such disadvantages are from the moral point of view arbitrary, and insofar as disease induces these disadvantages, then, in some theories, resources should be used to counter the effects of disease. Is this Brody's view? Does his view have moral premises of this sort?

One final comment is in order. We have, as we would all agree, major and unsettled social and institutional problems about a proper retirement age, especially if there is a mandatory age cap. If the limits of life expectancy are likely to exceed 85 years of age, what implication would this projected fact have for retirement planning, including a great deal of planning now being done in universities? Can we learn from the Japanese experience, where life expectancy is already longer? Is an implication of Brody's argument that we should protect ourselves against various well-known financial problems that threaten our future well-being by uncapping the mandatory retirement age (as the Reagan White House has recently mandated)?

Consider university professors, most of whom are Teachers' Insurance Annuity Association-College Retirement Equities Fund (TIAA–CREF) participants. 80 percent retire before they are 67 years old. Only 21 percent of TIAA–CREF retirees retire because they have reached the employer's mandatory retirement age, although another 7 percent are forced out by budget cuts, elimination of positions, and the like. But, still, 66 percent of university professors retire by choice at a time they select, and another 6 percent retire because they are disabled. For this large group of individuals uncapping the mandatory retirement age is now, in effect, an insignificant proposal; and the more attractive the retirement plan a university has, the higher the percentage of voluntary retirees will become. Perhaps these policies should be reversed, and instead of producing incentives to retire early, we should provide incentives to retire as late as possible.

But major problems also confront this policy alternative – particularly problems of how to maintain quality education for students, quality research, and fair employment principles for younger faculty or potential faculty, including those who represent groups discriminated against or formerly discriminated against. The young, the female, and the minority faculty have a great deal at stake in any decision about uncapping the mandatory retirement age. The further the mandatory retirement age is extended, the more difficult it will be for younger persons to find employment and to achieve tenure. Here we potentially encounter an enormously complicated set of trade-offs: In order not to retire a certain number of elderly faculty too early, we force out or avoid hiring younger faculty. Here one could argue age discrimination either way: The young are discriminated against because of their age levels, and so are the elderly.

It is the intractable and staggering character of such trade-offs that I have tried to point to in this comment.

Kennedy Institute of Ethics
Georgetown University
Washington, D.C.

ALBERT ROSENFELD

CHANGING IMAGES OF DEPENDENCY IN PROLONGEVITY

Images are 'only' images – mental constructs, mere abstractions. But they have concrete consequences. Our feelings and actions grow directly out of what we imagine to be true. Hence our images – though they may not initially conform to present reality – help shape ensuing reality. Today, our overriding image of the elderly is one of dependency. We expect them to be dependent, we perceive them as actually being dependent, and we behave toward them in accordance with this perception and this expectation. A number of authorities in the psychosocial aspects of aging – among them Robert N. Butler, Bernice Neugarten, Matilda Riley, and James E. Birren – have written of, and personally communicated to me, their conviction that many old people who would *not* otherwise fall into a state of dependency do so because all the signals we transmit to them, verbally and nonverbally, persuade them that they are not capable of being independent or nondependent (see, for example, [4], [7], pp. 1–21; [19], pp. 17–18, p. 29, p. 180, and [23]). As Sharon Curtin has put it ([20], p. 214), "The old are supposed to be a burden, not a resource." Similar convictions have been expressed by previous speakers on this program. Even now, then, before we enter the age of prolongevity, the image of dependency – its descriptive accuracy, as well as the wisdom or the usefulness of clinging to it – needs reappraisal.

When we think of old people in the future, it becomes even more important to change this dependency image, because its impact then will be not merely on individuals, but on how we look at our imagined future society, what we guess the nature of its problems to be, and the kind of solutions we arrive at. Yet, when we project that dependency image all the way into the next century, it still does not seem to change – it rather becomes magnified. Our demographic projections, especially with regard to the older segment of future populations, tend to be extrapolations from the present. The prospects are usually depicted as grim indeed, with an ever-increasing portion of the population growing older and older, and sicker and sicker, placing ever greater financial burdens on the younger groups that make up the work force ([1], [2], [4], [6], [7],

29

Stuart F. Spicker, Stanley R. Ingman, and Ian R. Lawson (eds.),
Ethical Dimensions of Geriatric Care, 29–44.
© *1987 by D. Reidel Publishing Company.*

[26]). Again in Curtin's words ([10], p. 195), "This is a culturally sanctioned attitude, sanctioned by the fear of those caught in the middle of the three-tiered society; those who must pay the bills, the middle-aged, working American. Goaded by fear of their own aging, angry at the financial drain caused by the dependent young and the dependent old, jealous of youth and hating the dreary future, they exist in a world of nameless dread and uncertainty. They feel as if they work for everyone's happiness but their own." Thus, in the future, the *quantity* of dependency is seen as rising, in terms of the sheer number of dependent individuals. Moreover, we envision the old women and men of the future, dubious beneficiaries of further biotechnological progress, as lasting for a stretched-out period of years, kept alive by heroic measures and artful medications, probably warehoused more and more in nursing homes, low in both vigor and functional capacity if not indeed steeped in total decrepitude – so that the *duration* of dependency also increases.

Though hope is held out that medical advances will continue to occur, with perhaps some significant breakthroughs in fields such as cancer and the cardiovascular diseases, on the whole one is led to envision an essentially infirm population of elderly, requiring long-term medical and custodial care, implying a budgetary crunch (already well under way) that may rival the nation's defense expenditures. (Walter McClure, in his presentation, underlined these dangers.) I know of no projections that forecast a population of non-elderly elderly, of aging but not aged individuals who remain in good, vigorous health until their most advanced years; I know of hardly any that predict breakthroughs in understanding the aging process itself, understanding that will allow stretches of prolongevity for which there is no human precedent. Yet, in the opinion of many leading gerontological investigators – including Roy Walford, Bernard Strehler, W. Donner Denckla, Leonard Hayflick, Alex Comfort, Denham Harman, Richard Cutler, and F. Marott Sinex, as well as leading Soviet investigators – such breakthroughs are likely to occur (see especially [8], p. 37; [12]; [20], Vol. A, pp. xiii–xv, pp. 8–15; and [27], pp. xi–xiv, pp. 161–175); and this scenario is, in my view, the more probable one to think about for the 21st century. We may well discover the basic mechanisms of senescence, enabling us to delay the aging process itself; and there will surely be many remedies offered (some are already in experimental use) for the prevention or alleviation of the specific symptoms of the disease we call aging. We could see not only an increased *average life expectancy*, but also, for the first time in human history, an extended *maximum life span*. As we eliminate or minimize the familiar ravages of aging – the aches and ills, the dysfunctions and incapacities, the stoop and shuffle, the sag and flab, that we have always observed to be the inevitable concomitant of time's passage – it becomes no longer preposterous to speculate upon the possible obsolescence of senescence.

To the extent that we ignore these possibilities in our projections of the future, to that extent do we risk that our ethical, economic, social, political and other considerations related to aging (and I mean related to a new kind of non-aging aging) will be unrealistic, if not useless. As one example, instead of worrying mainly about how the young will carry the financial burden of the old, a more primary concern may be how the young will find a place for themselves at all in a world full of still energetically active old people. The problem would thus center not on dependency but on *in*dependency. And among the appropriate solutions would surely have to be a recognition of *inter*dependency.

The Co-Directors of this symposium, in their letter of invitation to prospective participants, said they "do not believe that our society will witness a radical compression of morbidity and the end of infirmity among the aged at the turn of the century;" [Nor do I believe that we will see the *end* of infirmity, and certainly not quite that soon, but we are here talking about the 21st century, not just its 'turn'.] "On the contrary," they continued, "we anticipate that demographic and other data will forecast extensive late-life dependency." They were undoubtedly expressing the majority view, which could of course turn out to be accurate. Since I cannot be certain that the avant garde gerontologists will be as successful as they now imagine, I would be foolish to advocate that we *drop* the dependency scenario. But I am convinced that the grounds for their optimism are so sound that I think we would be ill advised to give short shrift to alternative scenarios in which a good measure of late-life non-dependency is a seriously-taken ingredient.

Because the avant garde view ([20], [22], [27]) has yet to become the prevailing stance among mainstream gerontologists – in fact still seems outlandish to many – it would be premature, at the least, to state flatly that old age, as we have always known it, is about to be abolished. But it seems safe to predict that, if we are to conjure up anything like accurate images of the elderly – at least the chronologically elderly – of the 21st century, we will have to call upon all the imaginative powers at our command. My own bets are already on the table ([22] pp. 204–219) that these images will not be images of dependency. The new images of non-dependency will not be so important in themselves as for their overall portents for society. There will be concern about the consequences of prolongevity in terms of exacerbating population and environmental-pollution problems, of the impact on retirement programs and social security benefits, insurance premiums, job availability, personal relationships, multiple careers, multiple marriages, character changes at the level of both individual and community, attitudes toward procreation, parenting, therapeutic options, inheritance, taxes, and death (though Benjamin Franklin predicted that we would never be able to do much about these latter two). Hardly an area of our lives will go unaffected, and some of the effects will be profound and far-reaching, if the nature of aging is radically altered and our life spans extended ([22], pp. 240–284). I do not plan to spend much time on such questions in this paper, simply because they are not the principal concerns of this volume. But clinical care of the aged *is* the central focus here – and our varying scenarios of dependency and independency do imply radically different clinical approaches.

Though, again quoting from the invitation letter, "this symposium is not devoted to the biological level of research which bears on the phenomenon of human aging," I feel a significant portion of this paper must be devoted to exactly that. For it is only fair now for the question to be asked: On what grounds do I take license to speak of a coming age of prolongevity? Is it truly a feasible alternative to the more dismal picture ordinarily presented when demographers forecast population trends over the coming decades?

Our traditional stance toward those who sought to overcome aging – a stance both justifiable and appropriate in the past – has always been one of derision, with a measure of pity reserved for the Ponce de Leons who vainly hunted fountains of youth and for the alchemists who brewed proleptic elixirs of life. Why, then, should we regard the gerontologists, with their new optimism, any less deluded or benighted than those earlier yearners for prolongevity? The legitimate grounds for their hopes lie in the vast quantities of new knowledge and understanding acquired over the past few decades about the basic nature of life and its detailed molecular mechanisms. The new facts and insights come from advances that can only be termed explosive in virtually all of the biomedical sciences. The contributions come from physiology, endocrinology, immunology, cell biology, the neurosciences, biochemistry, and especially from molecular genetics – as well as from an impressive array of new biotechnologies associated with these disciplines.

Even before this great outbreak of breakthroughs took place, it had already been convincingly demonstrated in numerous laboratory experiments that the built-in, species-specific life spans of many creatures are not necessarily immune to our intervention. As early as the 1930s, Clive McCay [16] showed that carefully-administered underfeeding (a semi-starvation diet, with calories cut by as much as two-thirds but nutrients maintained in good balance) of rats, begun before puberty, resulted in a considerable extension of their life spans. Only in the past few years, Roy Walford and his associates have shown ([27], pp. 98–113) that underfeeding can work for adult rats as well – and has in fact adopted 'undernutrition' as part of his own personal lifestyle. It has also been known for some time (though in this case none of the investigators has volunteered to try it personally) that when an old rat is joined to a young one via parabiosis (a surgical procedure that enables them to share a common blood circulation as Siamese twins do) the old rat lives longer than his unhooked-up littermates. When roaches are joined parabiotically, the old roach regains its youthful capacity to regenerate severed limbs. Animals have gained the benefits of prolongevity by eating antioxidants with their food, by having their immune systems manipulated, by having their temperatures lowered by a few degrees (this latter especially in cold-blooded creatures). Because calorie restriction works mainly at the beginning of life, while cooling mainly prolongs the latter part of life, Charles Barrows was able to triple the life span of rotifers. The species he used had a normal life span of 18 days. By restricting its food at the outset, he was able to extend its life to 36 days, adding a full life span. By cooling it toward the end of its life, he was able to add yet another 18 days.

P. L. Krohn discovered in the late 1950s and early 1960s that if skin cells from an aging mouse were transplanted to a young mouse of the same species, then re-transplanted to another young mouse as that one aged, and so on (though not indefinitely) in serial fashion, the transplanted cells would outlive their original donors by a wide margin. The same turned out to be true for human fibroblasts in tissue culture: with the addition to the culture of substances such as hydrocortisone or Vitamin E, the cells went on dividing longer than they otherwise would have.

W. Donner Denckla, who believes the pituitary releases a pro-aging hormone (or family of hormones), administered a crude pituitary extract to young rats, causing them to age prematurely. Conversely, by removing the pituitaries of older rats (thus presumably cutting off production of the aging hormone), then giving them replacement doses of thyroid and steroid hormones, was able to restore a number of youthful functions ranging from cardiovascular and immune parameters to DNA synthesis and fur growth. David Harrison has duplicated some of these feats in mice. (See [22] for longer accounts of all these experiments as well as additional references.)

And so on.

True, rats and roaches and rotifers are not the same as people. But people *are* living organisms, and gerontologists see no basic theoretical obstacle to adding good years to the human life span – or at least seeing to it that all the years of the current human life span are lived out vigorously and in good health until very close to the very end.

Though this may sound like sheer fantasy, remember that we are talking about the future – but probably the not-too-distant future. The achievement of this kind of prolongevity would require much lesser breakthroughs than we have seen occur in the last 10 to 20 years to bring about, for instance, the present state of the genetic-engineering art – which, if predicted, would have seemed wildly beyond the hope of attainment. Getting from Here to There in aging is much easier to believe, though of course an enormous credibility gap still exists.

Most theories of aging, at least in my limited taxonomy, break down into two rough categories: *wear-and-tear-theories*, wherein a given aspect of the wearing-away or running-down of the machine is deemed to be the primary cause of aging; and.*program* theories, in which a 'clock of aging' (if not more than one) is presumed to be genetically programmed into the organism.

A few examples of wear-and-tear theories:

Lipofuscin accumulation: Pathologic studies of cells, particularly nerve, liver, and heart-muscle cells, often show very large accumulations of lipofuscin, a fatty, pigmented, waxlike substance thought to represent undisposed-of cellular garbage. At advanced ages, lipofuscin may take up as much as 20% of the cell's space, significantly interfering with all its metabolic processes. Most gerontologists now consider lipofuscin more a consequence than a cause of aging ([24], pp. 252–264).

Cross-linkage: DNA, RNA, proteins (including enzymes, hormones, and the extracellular molecules, such as collagen and elastin, that make up the connective tissue), all the body's macro-molecules, in fact, are vulnerable to being cross-linked – being joined together in unintended ways. This happens in the normal hurly-burly of everyday metabolic life. The cell can correct much of this damage, but it loses some of this capacity as the body ages, and more and more cross-links become irreversible. The image that always comes to my mind is of two workers being handcuffed – or, worse still, lassoed so that each has both arms immobilized. Cross-linked molecules can be crippled in a number of ways, but they all spell decreased efficiency, at least – and (depending on which molecules are linked where) the total stoppage of a vital biochemical cycle in a cell, or a bottleneck on a critical assembly line (as the assembly of proteins from amino acids on the ribosomes), or the stiffening of connective tissue, or indeed the wholesale death of cells over time. The cross-linkage theory is mainly associated with the name of Johan Bjorksten, who feels his theory [5] has long ago been proven beyond a reasonable doubt. No one doubts that cross-linkage does take place and is an important element in aging, but the current consensus regards it as a manifestation rather than the initiating cause of aging.

Free radicals: In the course of the body's normal and necessary oxidation reactions, free radicals are often left over as highly reactive by-products. They are essentially broken-off pieces of molecules, most of which don't belong anywhere, but because each has a free electron yearning to unite with the first assailable molecule that comes along, it can create havoc in the neighborhood; as it slams in to join its victim molecule, it may create a small chain of other free-radical reactions before some kind of balance is restored. Free radicals have been shown to be heavily involved in radiation damage as well as other kinds of injury – for instance, the continuing damage to the heart muscle and nervous tissue after, say, a heart attack or a serious head or spinal injury. Free radicals are also considered to be a major cause of both cross-linkage and 'lipofuscin accumulation. On the whole, the free radical theory, first put forward by Denham Harman in the 1960's [13], has held up remarkably well and has continued to gain adherents over the years ([14], [22] pp. 45–47). It is my own favorite among wear-and-tear theories.

Somatic mutation: Genetic mutations as employed in evolutionary theory refer to changes in the DNA of the germ cells only. In this context, I am talking about mutations in the somatic cells – anything other than sperm or egg. Originally somatic mutation referred to 'point' mutations on a gene, the kind of change that can be caused by the random impact of, say, a cosmic ray or a potent chemical. When Leo Szilard [25] turned his attention from physics to biology in the 1950s, he hypothesized that random 'hits' of various kinds accumulate over the years to impair or destroy the functioning of genes, gradually causing the decline and death of cells and ultimately the decline and death of the whole organism. The somatic mutation theory was worked out in greater detail by Howard Curtis ([11], pp. 41–81) and others ([24], pp. 274–284). And of course it does make sense that a handicapped set of genes would adversely affect all the cells' – and hence all the body's – activities.

The genes have turned out to be much more complicated than we originally thought (that is, once we began to understand them at all). They coil and 'supercoil' into intricate configurations, and the irreversible unraveling of these structures is also now suspected of being instrumental to the aging process. In fact, this makes a theory of aging on its own. DNA is also now known to possess a built-in capacity to repair itself; and the *decline of DNA repair capability* constitutes yet another theory of aging ([20] Vol. 13, pp. 145–173) and one that does not want for advocates. I include these under the somatic mutation theory because they all involve changes of the genetic material in somatic cells, however they may be caused. Nor have I by any means, through this brief recitation, exhausted the ways in which these changes come about. One would expect, of course, that damage to some genes would be more critical than others. A mutation on a repeated sequence would probably cause no problem at all. A mutation on a regulatory gene might be much more menacing than one on a structural gene. One wouldn't want to see too many repair genes knocked out.

Error catastrophes: Once any living organism, including a woman or a man, is full grown, the main function of the cell appears to be the manufacture of proteins in order to rebuild its own structures, or, in some cells, for export as well – hormones, for example. In the course of protein synthesis, errors are occasionally made – perhaps by the DNA, one of the RNAs, the nuclear proteins and enzymes, the enzymes out in the cytoplasm, or any number of other molecules. The cell can usually rectify these errors; and, as a rule, the error occurs in a place that doesn't interfere with vital functions. But if the error goes uncorrected, and if it has impaired some crucial molecule, an indispensable job may fail to be done, and the result will be further error upon error – in a word, an 'error catastrophe.' Because, with aging, cells' self-repair capacity does decline, and the likelihood of such catastrophic occurrences does increase, Leslie Orgel came up with the error catastrophe theory of aging [18] – which, though it does not at this moment have many adherents, provoked an enormous amount of fruitful thinking and experiment at the time it was propounded.

The decline of immunity: The immunological theory of aging is generally attributed to Roy Walford [29], though Macfarlane Burnet is also one of its originators. As the immune system ages, at least two things happen: (1) immunity against infection goes down, and (2) autoimmune reactions go up. It appears that either senescent changes in the body's cells make them 'look foreign' to the immune cells, which then treat them as if they were invading microorganisms; or changes in the immune cells themselves impair their recognition abilities. More than sufficient reason, in either case, to connect immune decline with the aging process; but, again, the connection need not be causal.

As for 'program' or 'clock-of-aging' theories, there are two main contenders – and, in my opinion, they will turn out to be collaborative rather than competitive. Once there is agreement (among those who so agree) that aging is a genetically programmed sequence of events (though the precise timing is at least partially dependent on the organism's individual history in the world), the question left to argue is: Where is the clock?

One school holds that the clock is in the brain, and particularly in the hypothalamus and pituitary. The competing theory is that the clock is located in the genome of each individual cell. In either case, since the brain's release of hormones also requires genetic instructions transmitted from the nuclei of key cells, we are talking about a genetically programmed clock of aging. Here again, if one clock exists, it doesn't rule out the other. If space engineers have the wit to build fail-safe backup systems into their planetary probes, why would God or evolutionary nature have arrived at less clever designs? Nature, in any case, is nothing if not redundant. The clock theories:

The clock is in the cell: Leonard Hayflick was the first to demonstrate convincingly that cells in tissue culture have a finite life span [15]. Hayflick's WI-38 cell lines will double 50 times (plus or minus 10) before they quit; and they start showing internal signs of age before the doubling ends. If they are deepfrozen after 30 divisions, then thawed, they will only divide about another 20 times. By numerous and devious manipulations of these cells, their nuclei and their cytoplasms, Hayflick and his collaborators were able to prove that the controls for this 'aging under glass', as he calls it, reside in the nucleus of each individual cell. He is now trying to pinpoint these controls in the genome.

Richard Cutler, Roy Walford and Bernard Strehler are among those who believe there are good evolutionary reasons for postulating that the genes involved in aging are few in number ([20], p. 53, [24], p. 367) therefore worth seeking with some reasonable hope of success. Walford has been zeroing in on the major histocompatibility complex (MHC), a cluster of genes he thinks of as a kind of 'supergene,' all located on a single, known chromosome [28]. The MHC seems to be involved in various immune functions, including cell compatibility for transplants, as well as in DNA repair and some of the body's natural antioxidant systems that help combat free-radical damage.

The clock is in the brain: The major proponent of the most radical theory in this area is W. Donner Denckla [12]. To put his main idea in a capsule: Denckla believes that the pituitary begins, at puberty, to release a hormone or family of hormones that causes the body to decline in function at a roughly preprogrammed rate. He calls the hormone (not yet isolated and purified) DECO – for decreasing oxygen consumption – though others have referred to it as the 'aging' or 'death' hormone ([3], [12], [21]). One of its major mechanisms is to reduce or block the cells' ability to take in and utilize thyroxine, which is the principal *rate-controlling* hormone for the two main body systems – the cardiovascular and the immune – the failure of which represent the main causes of death. Denckla's theory is backed by many years of experimentation with thousands of rats. I have already described some of the manipulations that have enabled him to bring back youthful qualities and functions – including vital capacity, generally considered to be one of the few truly reliable indicators of aging – in old rats whose pituitaries have been removed.

Could Hayflick and Denckla both be right? They could, and I think they may well be. Questions naturally arise: If a hormonal clock, based in the brain, controls aging, then how could it possibly affect cells in tissue culture, outside the body and unreachable by any messages from the pituitary? The only answer seems to be that there is another clock, perhaps part of the genetic fail-safe system, resident in the nucleus. That nuclear clock would behave differently if it were still in the body, controlled and influenced to a greater or lesser extent by its natural physiological environment. The picture I like to use is that of, say, a violinist, or a whole string section, waiting for the conductor's cues before a bow is lifted or a note is played. In tissue culture, away from the conductor's influence, it could be that the string score is simply played out without interruption (since we are dealing here with mechanically programmed musicians, with all inhibitions removed). In any case, my view is that reasonable evidence exists to make a case for both clock theories. Walford, too, seems now ([27], pp. 172–173) to be advocating a 'two-clock model' of aging.

In discussing clock vs. wear-and-tear theories, a question that frequently arises is: Inasmuch as practically any one of the wear-and-tear theories, let alone all of them taken together, can account for just about everything we see happening in aging, why postulate a genetic program, a clock of aging, at all? (Alex Comfort once complained to me that almost any facet of the aging process can be parlayed into a viable theory of aging.) Who needs it? There are a number of answers, and I will just tick off a couple:

Throughout the realm of biology, all creatures seem to have evolved a species-specific life span. A shrew will live, say, a year and a half, while a Galapagos tortoise will go on for a century and a half, or even longer. If it was just a matter of wear and tear, would we not expect to see, just every now and then, a shrew that is 150 years old, and a Galapagos tortoise that dies of old age at one and a half? But we never do see that sort of phenomenon. (In humans, we do see rapidly accelerated aging in the progeroid syndromes, but these certainly seem to be genetic rather than wear-and-tear phenomena.) Or take cells growing in tissue culture. Normal cells have a finite life span; they age and die at roughly the 'Hayflick limit.' Cancer cells, on the other hand, go on and on indefinitely; they do not age and die. If a line of cancer cells is exposed to the same laboratory culture medium, the same chemical environment, the same conditions of wear and tear, as a line of normal cells, how is it that the normal ones age and the abnormal ones do not?

In all the diversity of aging theories (a standing cliché used to be that there are as many aging theories as there are gerontologists), there now does seem to be some convergence toward unification of a sort. For instance, if free-radical damage is what causes both cross-linkage and lipofuscin accumulation, as many believe, it doesn't replace those theories; it merely encompasses them. Or if programmed aging turns out to be valid, one can easily see how this could account for all the phenomena on which most of the other theories are based.

So much for theorizing. What does all this imply in terms of possible action scenarios – for therapies, preventive measures, cures, for the individual symptoms of senescence or for the aging process itself (which, if 'cured', would not necessarily cure all the damage already done)?

Let us suppose that Denckla were suddenly to come up with conclusive, utterly convincing, and universally accepted proof that he was right, and a decision made to proceed apace. It would be a matter of isolating, purifying and synthesizing the hormone (which would perhaps already have been done as part of the rock-hard proof), then perhaps fashioning some sort of anti-hormone – a long, tedious process, but not so long or tedious as it used to be, now that we have those incredibly automated protein sequenators and gene machines to accelerate the process. A shortcut might be available, though: We are now pretty sure that the pituitary – which we used to call the 'master gland' – cannot secrete any hormone without receiving from *its* master, the nearby hypothalamus, a specific command in the form of a 'releasing factor' (RF), a much smaller and easier-to-make molecule. It is also assumed that a similar-sized inhibitor molecule probably exists for every hormone. For the growth hormone, for instance, both the RF and the inhibitor have been found. If we possessed the hypothalamic inhibitor for DECO, could we stop the aging clock? It's not inconceivable. That is the way, after all, that one of the birth-control pills currently in use does work. That kind of success with DECO would appear to be still years away, and, if and when it does arrive, it will certainly require a long period of testing, for the side effects could be potent indeed. But a DECO inhibitor as an anti-aging medication can now be at least legitimately imagined, and not merely fantasized.

And what of the genetic clock in the cell nucleus? If it is there, we will presumably find it, perhaps in the MHC, as Walford suspects. Our genetic engineering capabilities have so far exceeded all forecasts that the removal and insertion of genes virtually at will, and not too far down the road at that, is so taken for granted as to occupy many a bioethics symposium with the dilemmas of when, if ever, the knowledge should be applied to the human genome. (In fact, three days from now, I will be participating in an intensive workshop in Washington on the priorities, challenges and dilemmas of gene therapy – sponsored by the Office of Technology Assessment of the U.S. Congress.) Tampering with a DNA-based clock should certainly present no insuperable *technical* difficulties.

Whether or not genetic aging clocks exist, and whether or not we discover them, a number of anti-aging substances are already being looked at (in fact, used – sometimes by the researchers themselves – though without hard evidence) as possible preventers, retarders or ameliorators of wear-and-tear damage ([20], pp. 135–147; [22], pp. 163–165). A host of antioxidants – among them Vitamins C and E, glutathione, beta carotene, and selenium – look promising indeed; there are good theoretical and epidemiological reasons for looking upon them as potentially protective against cancer as well. Among the immune boosters, the thymosins (thymic hormones) that are perhaps being pursued most aggressively by Allan L. Goldstein ([20], Vol. A, pp. 169–197) and his associates could, if they live up to their potential, considerably improve the quality of our later years by making us more resistant to infection and stress. There is another hormone, a steroid called DHEA (for dehydroepiandrosterone) – not only the most abundant steroid hormone in our bloodstream, but behaving like no other, in that its presence diminishes in a straight line with age. DHEA has only recently surfaced in our awareness (thanks largely to the research efforts of investigators like Arthur Schwartz and Douglas Coleman) as a promising anti-obesity, anti-diabetic, anti-cancer and anti-aging drug ([20], Vol. B, pp. 267–278).

Then there are lipofuscin scavengers and lysosome membrane stabilizers, as well as neurotransmitters and their precursors which, it is hoped, will hold off memory loss and other forms of cognitive and intellectual decline. There may be enzymes (to replace those that decline with age) and enzyme inhibitors (for those that are in oversupply). And of course there are the changes in lifestyle, including dietary controls. Roy Walford only in the past few years found a way to make McCay's early calorie-restriction techniques work effectively (though not so dramatically) in adult as well as in pre-adolescent animals – and he has volunteered himself as the first subject for his first human experiment ([27], pp. 98–113).

Though it would be fatuous and deceitful to claim that much has yet been *proven*, enough suggestive evidence has been coaxed forth to convince many gerontologists that there is an overall aging process, that

it is discoverable, and that we can do something about it. Most of our
current medical research is aimed at specific diseases; in the elderly, this
means cancer, heart disease, stroke, hypertension, arthritis, diabetes,
osteoporosis, the various forms of mental deterioration and dysfunction
(such as Alzheimer's disease), and all the rest. Most of our major
diseases seem to come in two forms – a juvenile or early-onset type, and
an adult or late-onset variety. The juvenile forms, which claim fewer
victims by far than the adult types, are believed to be mostly genetic or
familial in origin (though the opposite argument has been made for
diabetes); whereas the vast majority of cases come in later life and go up
steadily with age. In the lab, the same techniques that retard aging, such
as calorie restriction, also retard cancer and other late-onset diseases. It
would look as if the shortest route to the solution of many of these
diseases would be to do something about aging itself (after a coronary-
bypass operation, for instance, the atherosclerotic process just goes
right on as before). So, if something were to be done about the aging
process, in addition to seeing many fewer totally dependent patients in
the clinic, one should also see many fewer cases of all these deteriorative
diseases.

The title of this particular session has to do with prospects in the
coming 'prosthetic era.' I have deliberately avoided this phrase so far,
simply because I think that we can and will go beyond prosthetics.
Meanwhile, of course, we will see the continuing development of organ
transplant techniques, of organ assists (such as the pacemaker), of
artificial organs, and the like. We have already witnessed remarkable
success in the first experiments in the barely beginning field of fetal
brain grafts, which have already changed behavior and served as an
apparent cure for a rodent form of lab-induced Parkinsonism. There
were some early experiments, largely stopped now on ethical grounds,
in which electrodes were placed in the brains of patients who could push
the buttons on their own portable machines to stimulate the desired area
of the brain; this technique could be readily revived, with sufficient
reason.

I am by no means attempting to catalog all the possibilities exhaus-
tively (though I may be doing so exhaustingly) merely to give an
impressionistic overview of what may loom ahead that will affect how
specialists in geriatrics treat their future patients. If some of the specula-
tive eventualities I have touched on were indeed to come about, we
would need fewer specialists in geriatrics, since there would be a sharp

diminution in both the incidence and prevalence of senescent syndromes, and presumably much less disability to cope with. In people who lived a lot longer, it may be that dentistry would become a major problem. It could be that skin cells, which so far have not been called on to keep dividing beyond their capacity, would, in prolongevity, begin to have difficulty replacing themselves, providing unprecedented kinds of problems for dermatologists. How much deteriorative disease would remain, how common it would still be to find mental deficits among the elderly, is hard to foretell.

It is not my role here to try to spell out or even to visualize in detail the nature of clinical geriatrics in the 21st century. The message I have to deliver, and I feel I must do it forcefully, is that the route to prolongevity is one we can very feasibly hope to travel, and within reasonable time frames; success of whatever magnitude will create a diversity of options, implying a diversity of treatment scenarios. We have all come with a clear resolve to get as good a fix on the future as we can, to help ourselves and others think about what the practice of clinical geriatrics will be like in the 21st century. To that end, I urge that we do not handicap ourselves by limiting our considerations to the circumstances that are usually forecast, but which, in my opinion, are not those most likely to come about.

University of Texas Medical Branch
Galveston, Texas

BIBLIOGRAPHY

1. Asimov, I.: 1975, 'The Coming Age of Age', *Prism* (Jan.), 52–56.
2. Bevan, W.: 1972, 'On Growing Old in America', *Science* (Sept. 8), 840.
3. Bilder, G. E. and Denckla, W. D.: 1977, 'Restoration of Ability to Reject Xenografts and Clear Carbon After Hypophysectomy of Adult Rats', *Mechanisms of Aging and Development* **6**, 153–163.
4. Birren, J. E.: 1960, 'Psychological Aspects of Aging', *Annual Review of Psychology*, 161–198.
5. Bjorksten, J.: 1968, 'The Crosslinkage Theory of Aging', *Journal of the American Geriatrics Society* (April), 408–427.
6. Budd, J. H.: Apr. 1, 1974, 'Opening Remarks', at Congress III: The Later Years, American Medical Association, Chicago, Ill.
7. Butler, R. N.: 1975, *Why Survive? – Being Old in America*, Harper & Row, New York, N.Y.
8. Chebotarev, D. F. (ed.): 1972, *The Main Problems of Soviet Gerontology*, Kiev, U.S.S.R.

9. Comfort, A.: 1967, 'On Gerontophobia', *Medical Opinion and Review* (Sept.), 30–37.
10. Curtin, S. R.: 1973, *Nobody Ever Died of Old Age*, Atlantic Monthly Press – Little, Brown, Boston, Mass.
11. Curtis, H. J.: 1966, *Biological Mechanisms of Aging*, Charles C. Thomas, Springfield, Ill.
12. Denckla, W. D.: 1981, Interview in *Omni* (Nov.), 91–95, 132–136.
13. Harman, D.: 1969, 'Prolongation of Life: Role of Free Radical Reactions in Aging', *Journal of the American Geriatrics Society* (Aug.), 721–735.
14. Harman, D.: 1983: 'Free Radicals and the Origination, Evolution and Present Status of the Free Radical Theory of Aging', presented at 13th Annual Meeting, American Aging Association (Oct. 8), Washington, D.C.
15. Hayflick, L.: 1970, 'Aging Under Glass', *Experimental Gerontology* (Dec.), 291–303.
16. McCay, C. M., Maynard, L. A., *et al.*: 1939, 'Retarded Growth, Life Span, Ultimate Body Size and Age Changes in the Albino Rat after Feeding Diets Restricted in Calories', *Journal of Nutrition* **18**, 1–13.
17. Neugarten, B. L.: 1971 'Grow Old Along with Me! The Best Is Yet to Be', *Psychology Today* (Dec.), 45–48, 79–81.
18. Orgel, L. E.: 1973, 'Ageing of Clones of Mammalian Cells', *Nature* (June 22), 441–445.
19. Pelletier, K. R.: 1981, *Longevity: Fulfilling Our Biological Potential*, Delacorte Press, New York.
20. Regelson, W. and Sinex, F. M. (eds.): 1983, *Intervention in the Aging Process* (2 vols.), Alan R. Liss, Inc., New York.
21. Regelson, W.: 1983, 'The Evidence for Pituitary and Thyroid Control of Aging: Is Age Reversal a Myth or Reality? The Search for a "Death Hormone"', in Regelson, W. and Sinex, F. M. (eds.) *Intervention in the Aging Process*, Vol. B, pp. 3–52.
22. Rosenfeld, A.: 1976, *Prolongevity*, Alfred A. Knopf, Inc., New York, N.Y. (Revised edition in press.)
23. Rudolph, M.: 1970, 'Sense and Nonsense About Growing Older', *Family Health* (March), 28–30.
24. Strehler, B. L.: 1977, *Time, Cells, and Aging*, 2nd ed., Academic Press, New York.
25. Szilard, L.: 1959, 'On the Nature of the Aging Process', *Proceedings of the National Academy of Sciences* **45**, 30–45.
26. *Time* (anon.): 1970, 'Growing Old in America: The Unwanted Generation' (Aug. 3), 49–54.
27. Walford, R. L.: 1983, *Maximum Life Span*, W. W. Norton, New York.
28. Walford, R. L.: 1983, 'Supergenes: Histocompatibility; Immunologic and Other Parameters in Aging', in Regelson, W. and Sinex, F. M. (eds.), *Intervention in the Aging Process*, Vol. B, pp. 53–68.
29. Walford, R. L.: 1969, *The Immunologic Theory of Aging*, Williams & Wilkins, Baltimore, Md.
30. Williams, J. R.: 1983, 'Alterations in DNA/Chromatin Structure During Aging', in Regelson, W. and Sinex, F. M. (eds.), *Intervention in the Aging Process*, Vol. B, pp. 145–153.

RANDOLPH MARTIN NESSE

AN EVOLUTIONARY PERSPECTIVE ON SENESCENCE

At the heart of gerontology, there is an important scientific problem that is now ripe for solution. The problem is: "Why does the phenomenon of senescence exist?" Not why, in the sense of the mechanisms that cause damage. Gerontology has studied dozens of such mechanisms and the resulting mass of information has proved most difficult to integrate. The question to be addressed here is not *what* the mechanisms are but, instead, *how* they have come to exist. In organisms shaped by natural selection, why is there aging and why do individuals inevitably die by some specific age? These crucial questions have not been systematically considered. An answer to the question of how natural selection has affected the traits that influence aging would bring together many current findings of gerontology and would offer a new perspective on what aging really is and its place in the pattern of life. An evolutionary understanding of aging may also shed new light on related ethical and personal issues.

PROXIMATE *vs.* EVOLUTIONARY EXPLANATIONS

Two different kinds of explanations are needed in order to understand fully a biological phenomenon like aging. One kind is the proximate explanation, and the other is the evolutionary explanation ([20], pp. 67–76). This distinction requires elaboration with a few examples. In order to explain any biological phenomenon, we must, of course, give a proximate explanation of the mechanism and how it works. The proximate explanation of the heart, for example, must include its structure, how the valves work, how the contraction is coordinated and regulated, and how the heart rate is controlled. The proximate explanation must include the details of the mechanism at every level. In addition to this proximate explanation, however, a separate, evolutionary explanation is necessary. We must also explain the adaptive function of the heart and the natural selection forces that have shaped these mechanisms. It is not hard to see the function of the heart – it circulates the blood so that each cell is nourished and wastes are removed.

45

Stuart F. Spicker, Stanley R. Ingman, and Ian R. Lawson (eds.),
Ethical Dimensions of Geriatric Care, 45–64.
© *1987 by D. Reidel Publishing Company.*

As a second example of these two kinds of explanations, let us consider the momentary glow of a firefly. The *proximate* explanation includes the anatomic and neurochemical mechanism that mediates and controls the flash of light. It also includes the developmental process that begins with a DNA code and results in the specific structures of the mechanism. The *evolutionary* explanation, in contrast, specifies the adaptive function that the capacity serves, the selective forces that have shaped it, and, insofar as it is possible, its phylogenetic history. The proximate explanation accounts for the workings of the mechanism; the evolutionary explanation accounts for its existence as a result of natural selection. It is quite possible to formulate testable hypotheses regarding evolutionary function. Does the firefly's flash serve to frighten predators or to locate food? No, both of these hypotheses are false. The function of the glow is to locate mates. A firefly that lacks this trait will survive, but it will not be likely to reproduce, so its genes will be lost.

The cough reflex is a third example. The proximate explanation includes the details of the anatomy and function of the sensory nerves from the respiratory tract, the brain's processing of neural impulses, the motor nerves from the brain to the respiratory tract and diaphragm, and the mechanism by which neural impulses influence muscular action. The evolutionary explanation is that this reflex functions to clear foreign substances from the respiratory tract, and this decreases the likelihood of disease and death. Individuals with an intact cough reflex have a selective advantage over those who do not, and the genes that code for this trait have therefore spread and become universal. Of course, many genes are involved, and natural selection actually shaped the cough reflex gradually over millions of years, not all at once.

The distinction between proximate and evolutionary explanations is well accepted, but some people remain concerned that evolutionary explanations based on the function of a trait are somehow not really a part of science. Biologists, too, were at one time suspicious of questions involving function. The victories over vitalism and teleology were not yet secure, and reductionism and imitation of the methods of physics were the order of the day. Now, however, consideration of a trait's adaptive function is required in every area of biology. The way was paved first by anatomists and physiologists, and then by Tinbergen and other ethologists, who had to account for patterns of behavior observed in various species [27]. More recently, Ernst Mayr has argued, in his book *The Growth of Biological Thought*, that this distinction defines two separate biologies:

The two biologies that are concerned with the two kinds of causation are remarkably self-contained. Proximate causes relate to the functions of an organism and its parts, as well as its development, from functional morphology down to biochemistry. Evolutionary, or historical, or ultimate causes, on the other hand, attempt to explain why an organism, in contrast to inanimate objects, has two different sets of causes, because organisms have a genetic program . . . ([20], p. 68).

No biological problem is fully solved until both the proximate *and* the evolutionary causation has been elucidated. Furthermore, the study of evolutionary causes is as legitimate a part of biology as is the study of the usually physical–chemical proximate causes ([20], p. 73).

The importance and legitimacy of the distinction between proximate and evolutionary explanations are steadily better appreciated by scientists, although interesting questions remain about this distinction, and the methods appropriate for testing evolutionary hypotheses. Space limitations preclude further discussion of these issues here. Instead, the distinction will be used to better understand the phenomena of senescence.

To date, almost all research on aging has been designed to provide proximate explanations. Textbooks describe the many changes that occur with age, the dozens of mechanisms that may be responsible, and evidence for and against each of these mechanisms [9]. The diversity and number of these facts and theories are a significant fact in its own right. Limits to cell division may play a role [14]. There may be errors in DNA replication, damage by free radicals, damage from the immune system, irreversible protein cross-linkages – the list goes on and on ([6], [22]). Not only is there no straightforward way to decide what contribution is made by each mechanism, it is not even clear which are mutually exclusive possibilities and which may make overlapping contributions. Something is missing here. An evolutionary perspective may provide an important theoretical framework.

There are four possible relationships between natural selection and aging. First, there is the possibility that genes have nothing to do with aging, and natural selection is, therefore, irrelevant. Second, it could be that aging is somehow adaptive, and has been selected like any other trait. Third, it is possible that the genes that induce aging have never been exposed to natural selection because animals in the wild never live long enough for the genes to pose any serious disadvantage. Finally, there is the intriguing fourth possibility that the same genes that are responsible for the problems of aging also have beneficial effects earlier in life, so that they are, therefore, selected and maintained in the gene pool. These four possibilities will be considered one at a time.

48 RANDOLPH MARTIN NESSE

WEAR AND TEAR

Many people think of aging as "wearing out." It is true that parts of the body do wear out and that this contributes to aging. Some parts, however, never wear out or are continually replaced. We never run out of red blood cells, for instance; they are continually replaced. Teeth wear out, but they could be regularly replaced since that happens once already for every human. A lizard can regrow its whole tail if necessary. The effect of aging is manifested not in wear itself, but in the body's limited and diminishing capacity to protect and replace its parts. These capacities, or lacks of capacities, are determined by the genes. Wear and tear cannot explain aging. Both protective and detrimental genetic factors are involved in aging, and we must offer an evolutionary explanation for their presence.

AGING AS AN ADAPTATION

The second theory considers aging to be an adaptation. Is aging itself somehow useful? This idea appears first in an 1881 article by August Weisman:

Worn out individuals are not only valueless to the species, but they are even harmful, for they take the place of those which are sound. Hence, by the operation of natural selection, the life of our hypothetically immortal individual will be shortened by the amount which was useless to the species ([28], p. 24).

There are many variations on this idea [24], but all of them propose that the individual ages and dies for the sake of the group or the species. This mechanism of group selection was once accepted by many biologists in order to account for phenomena that seemed to have no other possible explanation. However, William Hamilton advanced evolutionary theory considerably in 1969 when he formulated the principle of kin-selection to account for the seemingly self-sacrificing behaviors that had previously been used as prime examples of traits thought to have been selected as a result of their benefit for the group or the species [13]. It is not possible to pursue this interesting story in more detail here, except to say that kin-selection does not explain senescence in general, and that group selection has now been discounted as an evolutionary mechanism, except in very special situations. The problem with a theory that proposes an adaptive value for aging is quickly clear if we imagine an evolutionary competition between an individual who ages and dies, with an individual who does not age at all. If the individual who does not age lives longer and has more offspring, then the individual that ages, and that individual's offspring, will have relatively fewer of their genes represented in subsequent generations. This process would systematically eliminate genes that caused aging, even if they benefitted the group or the species. Aging is not an adaptive trait in itself. This second possible explanation for aging is as wrong as the first.

SELECTIVE IRRELEVANCE

The third explanation is based on the idea that the genes involved with aging are beyond the reach of natural selection because their effects never have any real disadvantage for individuals in the wild. Even if there were no senescence, a certain proportion of individuals would be killed each year by accidents, predators, disease, starvation, and other forces. Depending on the species and the conditions that it encounters, the mortality rate may be 3% per year or may be 50% per year. Whatever the mortality rate is, after some specific number of years there will be no individuals left. Any genetic effect that poses a disadvantage only after this age will be beyond the reach of natural selection. When individuals of this species are placed in a protected setting, they will live beyond the age that most live to in the wild, and the effects of aging will be seen. Death may then inevitably occur by a certain age because the body can no longer maintain homeostasis during the slightest stress. The core of this proposal is the accumulation of deleterious mutations with effects later in life, because selection cannot eliminate them.

This theory was first proposed by J. B. S. Haldane in his 1942 book, *New Paths in Genetics* [12]. "In man there is good evidence that arteriosclerosis and some other senile diseases are largely genetically determined. It is natural that such genes should accumulate as a result of mutation, for there is no selection against genes which act after the reproductive period" ([12], p. 113). Haldane deserves credit for originating the idea, but there are two problems that come from his focus on the end of reproduction instead of the inevitable demise of a certain number of individuals each year. First, any species that provides care or resources for its young will be subject to selection until the period of contribution to the offspring is over. A second more basic error is that Haldane assumes that an age limit to reproduction is inevitable. It is not. If reproduction ceases after a certain age, this must either be because changes of aging have disrupted the reproductive capacity or because the limit to the period of personal reproduction is somehow adaptive. This second possibility will be considered later. First, the theory of mutation accumulation must be illustrated. If there were a mutation that, for instance, caused steady, very slow deposition of calcium in the skin so that the skin became brittle by age 500, then there would be no direct action of selection to eliminate this gene because all individuals are dead from other causes long before this age. If another gene caused clouding of the cornea, but the resulting poor vision never became a problem until after the age at which essentially all humans had died when they lived under natural conditions, then selection could not eliminate or modify this gene. This theory of mutation accumulation is probably accepted by more gerontologists than any other evolutionary theory of aging.

Many give credit for the idea to P. B. Medawar who elaborated it in his 1946 article, 'Old Age and Natural Death' ([21], pp. 17–43). He claims in this article that he is only adding "a few extra guesses woven in among Weisman's original hypothesis of aging" ([21], p. 40), but this is far from the case. Medawar correctly notes the importance of the smaller number of old individuals and its relationship to the reduced force of selection. In this article, he clearly anticipates the pleiotropic theory, the fourth possibility considered here:

It is by no means difficult to imagine a genetic endowment which can favor young animals only at the expense of their elders; or rather at their own expense when they, themselves, grow old. A gene or combination of genes that promotes this state of affairs will, under certain numerically definable conditions, spread through a population simply because the younger animals it favors have, as a group, a relatively large contribution to make to the ancestry of the future population ([21], p. 38).

He fails to develop this idea, however, until his 1952 article, 'An Unsolved Problem in Biology' ([21], pp. 44–70). Even here, however, he instead emphasizes the fact that natural selection will tend, as the result of modifier genes, to push a deleterious genetic effect to later and later expression in the life cycle. He clearly delineates the appropriate method for recognizing the effect of senescence on a wild population. He points out that if organisms at each age are subject to the exact same rate of mortality, then the survival curve will show an exponential decline and a semi-logarithmic plot of survivors versus age will be a straight line. If, however, older individuals are more subject to predation and other dangers as a result of senescent changes, then the rate of mortality will increase with increasing age, and the semi-log plot will instead be a downward, sloping curve. Medawar notes the lack of life-table evidence for wild populations, then, without more ado, states that senescence is not observed in nature but only in laboratory settings. He apparently bases this on the absence of observations of decrepit animals in the wild, even though his own thinking clearly emphasizes that this is by no means the crucial datum on which to make a decision about the presence of senescence in wild populations.

Alex Comfort adopted this position in his enormously influential book, *The Biology of Senescence*, now in its third edition [4].

Death from senescence is itself in many species so rare in the wild state that failure to senesce early, or at all, has little value from the point of view of survival. In many forms the cessation or reduction of group breeding capacity happens well before senescence proper – with certain exceptions in social animals. What happens later, in the postreproductive period, is theoretically outside the reach of selection, and irrelevant to it ([4], p. 96).

There are two problems here. First, there is no explanation for the cessation of reproduction; an effect of senescence is presumed to be its cause. Second, when Comfort says "death from senescence," he apparently means natural death, but this is, again, not the issue. The issue is whether declining fitness makes an organism increasingly vulnerable to death from any number of causes. Few people have ever seen a very old rabbit in the wild, but this does not mean that aging is unimportant for them. If a one-year-old rabbit can run just slightly faster than a fox and a two-year-old rabbit can run just slightly slower, then foxes will catch many more two-year-old rabbits than one-year-old rabbits, aging will be a major cause of death, and it will be subject to a strong effect of natural selection. If this is the case, and for many species it may well be, then natural selection is acting strongly on aging, and the mutation accumulation theory is an insufficient explanation for the existence of genes that cause aging.

What actual evidence of mortality rates at different ages for a variety of species is available? Remarkably little. In an era that spends millions on research in basic science and millions, for that matter, on aging research, we still do not have field data on the mortality rates at

different ages for more than a few species. Gathering this data should be
a high priority because it will indicate how strongly selection is working
on aging, and, therefore, the plausibility of the mutation accumulation
theory of aging.

Some data are available. The easiest to interpret is that which shows
the rate of mortality per unit time, at different periods in the life span.
For humans, the data are good, at least for modern times. It shows that
the force of mortality increases from age two on, and increases exponentially, doubling every eight years, starting at about age 30 ([10], pp.
28–29). Senescence is occurring steadily and causing increased susceptibility to death, even in the '30s and '40s. A more comprehensive view of
aging in a given species is provided by the life table – a summary of the
number of surviving individuals at any given age. If there were no
senescence, then the curve should show a steady decline of a certain
percent of individuals each year. When senescence is present, the curve
becomes progressively more rectangular.

Systematic consideration of the importance of senescence in wild
populations requires a quantitative assessment of its strength in various
species, but such techniques have not been available. The effect of
senescence acting on a wild population can, however, be quantified as
the coefficient of selection that acts on the traits comprising senescence.
The arithmetic is very simple. Treating each sex separately, the mortality rate is determined during the period of early maximum reproduction.
The actual survivorship curve is then compared to a hypothetical one
which is based on the assumption that the force of mortality does not
increase at all with increasing age, that is, that there is no senescence. If
one further assumes, as is generally correct if there is no senescence,
that the reproductive rate remains constant with increasing age, then the
number of reproductive life-years lost to individuals who senesce, as
compared with those who hypothetically do not, can be readily identified. The coefficient of selection acting on senescence then can be seen
to be equal to the decrement in reproductive life years caused by
senescence, divided by the total number of reproductive life years for a
population which does not senesce, but instead loses members at a
steady rate. This technique will be described more comprehensively
elsewhere, but this simplified version shows the potential for comparing
the effects of senescence on a variety of species. A sample calculation
based on data published for the red deer population on the Isle of Rhum

[5] shows a very high coefficient of selection of 0.65. Collection and analysis of data for a variety of species would be of great benefit.

In 1957, Gertzing summarized the available data on senescence in wild fish populations [11]. This is of particular interest because many fish continue to grow and to increase their reproductive capacity throughout their lifespan, so they would be less likely than other species to demonstrate senescence. Nonetheless, his data show clear-cut evidence of increasing mortality with increasing age in several wild fish populations that had not been exploited by man. Data for small birds, on the other hand, seem to show a very rapid, steady mortality that is uninfluenced by senescence, so far as we can tell with the available data ([4], pp. 141–142). For tsetse flies, there is excellent evidence for the effect of senescence on mortality rates in the wild during the rainy season, but not during the dry season [18]. In plants, the importance of senescence is substantial [19]. Unfortunately, data of this sort are scarce, and what is available has not been analyzed from this point of view. It seems that scientists were so convinced that aging is not a factor for wild populations, that many opportunities to answer the question have been missed. A small institute, say, ten people funded for ten years, would take us a long way toward answering this very important question.

Separate from the problems of the available life-table data, there are also theoretical problems with the mutation accumulation theory. First, there is a problem of how the genes that affect aging could spread and become universal in the gene pool if they were indeed irrelevant to survival and reproduction. If they were not increased in frequency by natural selection, then how did they spread? The concept of genetic drift ([31], [3]) is a possibility, but it seems hard to imagine that this could account for the large number of genes that are involved in senescence and their remarkably uniform effects, both within members of the same species and between closely related species. Although the importance of drift is the subject of a complex technical debate, that cannot be pursued here.

The theory of mutation accumulation faces still other problems related to basic observations about aging and life span. If aging genes have not been subject to selection, one might expect that the effects of aging would be substantially different in different people, that different effects would become manifest at different rates, and that the length of life in a protected environment would vary considerably. What we find, however, is that the effects of aging are very similar in different individuals, that the rate of decline in reserve capacity in a variety of organ systems is identical [26], and that length of life for many species in a protected environment varies by only a small amount. Fries and Crapo have extrapolated current human mortality trends to estimate that, if premature causes of death were eliminated, 95% of people would die within 8 years of age 85 ([10], p. 71).

Finally, proximate studies on senescence challenge the idea that selection has not affected the genes which influence aging. Proximate research has vividly demonstrated the way in which the effectiveness of a variety of defenses against aging is correlated with the life spans of diverse species. The ability to repair DNA, the level of protection against damage from superoxides, the number of cell divisions possible – all of these show substantial correlations with the life span of the species considered ([6], pp. 45–57). The proponents of these theories have used this data to argue that each of these mechanisms contributes to the effects of aging. For laboratory studies, this may be correct, but an evolutionary perspective suggests that these protective mechanisms have been shaped by natural selection to be just as effective as they must be to protect these species during their usual life spans in the wild. They cannot provide more protection, because natural selection has no effect at ages not encountered in the wild. In combination with the other arguments advanced above, these facts pose serious problems for the hypothesis that aging is caused by the effects of genetic mutations that have accumulated outside the range of natural selection.

Although the accumulation of mutations is probably incorrect, the importance of the decreased force of selection with increasing age remains crucial. To go further, however, we must distinguish among three categories of genes. Each will be differently affected by natural selection. First, there are genes that protect against or repair inevitable damage at the molecular level. Genes that code for DNA repair mechanisms are a good example. Second, there are genes which code for mechanisms that allow the regeneration of damaged cells and tissues. The ability to heal a skin laceration, and the ability of a starfish to regenerate appendages are good examples. Third, there are genes whose effects cause tissue damage. Using these categories, it is readily apparent that the declining force of selection with increasing age cannot account for those genes which cause damage, but it can readily explain why the effects of various mechanisms that protect the body at the molecular and cellular level may not prove effective beyond the usual life span. With this perspective, it appears that wear and tear can cause aging because selection has not been able to create mechanisms that protect the individual long enough, or because the creation and maintenance of such protective mechanisms exacts a continuing cost to the organism which is not worth the benefit of improving the protective mechanism, given the limited life span of a species in the wild.

The limited ability of the body to regenerate damaged tissues is a somewhat different issue. Such limitations are imposed partly by the rarity of opportunities to repair some specific forms of damage. For instance, it would be extremely rare for an individual to live very long after trauma which damaged brain or heart tissue, so that the capacity to regenerate these tissues would offer little survival advantage. In addition, the benefits of complex organizations of tissue that can be achieved most efficiently with a single irreversible and unrepeatable process of differentiation may outweigh the costs of being unable to regenerate some tissues. Finally, the ability to regenerate damaged tissue must involve some risk of uncontrolled cellular replication. This risk of cancer may also be a cost that limits the selection force for mechanisms that allow regeneration of damaged tissue.

Two of the three categories of genes that influence aging can, therefore, be understood without too much difficulty. Natural selection has shaped mechanisms that protect cells and that repair damaged tissues, but these mechanisms cannot be perfectly efficient, both because of compromises that must be struck with inevitable costs, and because some of the damaging events are either rare or occur after the age at which essentially all individuals of the species have died in the wild environment. The third category of genes, those whose effects cause damage themselves, cannot be explained by the evolutionary mechanisms outlined so far.

THE PLEIOTROPIC THEORY

The fourth evolutionary theory of aging is usually called the pleiotropic theory. It was first formally stated by George Williams in 1957 [29]. He emphasized the decline in numbers of individuals with advancing age, even in the absence of senescence, and then he pointed out that the larger number of young individuals have, by the simple fact of their number, many more genes on which selection can act, than the smaller group of old people. Natural selection acts more strongly on genetic effects that are expressed in youth than it does on those effects expressed in old age, simply because there are more individuals for it to act upon. Pleiotropy refers to the fact that a single gene may have many different manifestations, and, in this case, that these are likely to be different, or to have different significance, at different ages. The idea is similar to that advanced by Haldane and Medawar, but Williams states it much more clearly, recognizes its central importance, and draws several interesting inferences.

An example offered by Williams will illustrate the theory. If there were a hypothetical gene that made bones stronger in early adulthood, this would offer a selective advantage, and the gene would spread in a population and become nearly universal. If this same gene caused steady deposition of calcium in the arteries, so that some people had strokes or heart attacks, even during the life span observed under natural conditions, then that gene would also pose a serious disadvantage. If the disadvantage were equal to the advantage – say, for instance, that the stronger bones increase survival by one percent per year in early adulthood, and the arteriosclerosis resulted in the death of one percent of older people per year, then the gene could still be selected for and could spread in the gene pool because there are so many more young people than old people for selection to act on. It is a simple, but brilliant idea. A gene with an early advantage will be selected, even if it causes a serious disadvantage or death at a later age. This theory, in contrast to the other three, can explain, in evolutionary terms, the existence of genes which cause tissue damage associated with old age.

A few more hypothetical examples will further illustrate the theory and its potential importance. If an individual with especially delicate structures in the lung could transport oxygen and carbon dioxide more rapidly, this would offer a significant advantage in the ability to flee from predators or to run after prey. If this same trait resulted in more fragile tissues that were more susceptible to damage and less susceptible to repair, this would pose a significant long-term disadvantage and might well mean that, after some age, the lungs could not be expected to function well enough to support life, even in the resting state. Even so, this gene would be selected and would become a part of every individual because those individuals who had it would survive better and reproduce more in youth (at a time when there are more individuals alive), while natural selection would not act nearly as strongly on the deleterious effects later in life. An individual who did not have this trait would contribute fewer of his genes to the next generation than an individual who had the trait, and the trait would become nearly universal. Whether this proposal is correct or not, it should not be surprising that pulmonary function steadily declines with increasing age.

As another example, let us consider a hypothetical mutation that resulted in a larger milk supply from the breast that would allow more offspring to survive in periods of famine. Even if this same gene caused a tendency to have breast cancer later in life, the gene might well be incorporated in the gene pool by natural selection. As a final example, consider the advantages that a particularly aggressive immune system would offer against possibly lethal infections. A gene which made such an aggressive immune system possible would be selected, even if it resulted in autoimmune damage that accumulated steadily with age.

William Hamilton has mathematically analyzed the pleiotropic theory of senescence [14]. After considering mathematical models for natural selection of pleiotropic genes, he concludes that "for organisms that reproduce repeatedly, senescence is to be expected as an inevitable consequence of the working of natural selection" ([14], p. 26). Pleiotropic effects must account for some of senescence. The pleiotropic theory avoids the problems of the other theories and accounts for phenomena which other theories cannot. It explains how there can be genes that cause the changes of aging, it explains why their effects are in synchrony, why mortality rates increase exponentially, and why the maximum life span is so rigidly fixed in many species.

Some other proposed theories of aging are tacitly based on pleiotropy. For instance, it has been suggested that limits to the number of fibroblast replications may serve the adaptive function of limiting the growth of atherosclerotic plaques. Although this proposal is unlikely, it is more clear when recognized as a possible pleiotropic mechanism to explain limits to cell division.

In his discussion of the pleiotropic theory, Williams proposes an ingenious explanation for menopause ([29], pp. 407–408). At first glance, it appears that cessation of reproduction must be a manifestation of senescence; it certainly seems to pose a disadvantage for maximizing the number of one's genes in the next generation. However, Williams notes that, in species that offer parental care to their young, selection continues to operate so long as this care is provided. After all, the offspring have many genes identical to the parents. He proposes that the senescent changes associated with increasing age make child-bearing more risky, and this threatens not only the life of the mother but also her ability to care for her existing offspring who carry replicas of her own genes. Thus, women who have menopause might have a selective advantage because more of their offspring will grow to reproductive maturity as compared with those mothers who continue to have children and risk the survival of the children already born. This proposal is difficult to test by the comparative method, because the duration of child care in humans is so much longer than in other species. Nonetheless, the possible evolutionary benefits of menopause may be of importance, especially to women and to gynecologists.

Are there specific data that support the pleiotropic theory of aging? We have already seen life-table data that show that aging is a factor in the mortality of wild populations. If further studies show that it is a major factor, that is, that the coefficient of selection is high in a variety of species, then this cannot be explained by the other theories. Once again, important data are not available.

The other main support for the pleiotropic theory of aging comes from breeding studies. If one allows fruitflies to breed only at advanced ages, then presumably this procedure will select against pleiotropic genes that tend to shorten the life span. This is exactly the experiment done by Rose and Charlesworth [23]. They found that this selection process increased longevity and late reproductive output, but that it decreased early reproductive output and decreased total reproductive output. They conclude, "It seems reasonable to suggest that senescence in *Drosophila* is due to the late-acting deleterious effects of genes which are favored by natural selection because of beneficial effects at early ages" ([23], p. 142).

Sokal performed the opposite experiment [25]. He bred 40 generations of flour beetles from eggs laid very early in the life span. He found that this breeding procedure produced significantly shorter life spans, and he concluded that this resulted from either the accumulation of mutations or from selection from pleiotropic genes. If he had measured changes in reproductive capacity, it might be possible to better use his data to support the pleiotropic theory.

A full consideration of the evidence that bears on the pleiotropic theory of aging is beyond the scope of this presentation. It should be clear, however, that it is a sensible theory, and that it has some experimental, as well as theoretical support.

The importance of pleiotropic effects will differ substantially for different species. A phylogenetic perspective offers intriguing predictions. If a species has recently, say in the past million years, been exposed to increasing predation or competition for food and shelter, this would decrease the average life span, and one would expect to find relatively few genes directly influencing this shorter life span in the wild, most of which would likely be pleiotropic. For such a species, average and maximum life span in a protected setting should show moderate variation. If, however, mean lifespan in the wild has recently increased because a species has been released from predation and competition, as appears to be the case for man, then one would expect to find a large number of pleiotropic and other senescent effects clustered tightly together by natural selection in a brief period at the end of life. Selection for increased efficiency of various protection mechanisms should be proceeding. There should be selection against pleiotropic genes, and this might cause susceptibility to diseases or other decreases in vitality early in the life span. For such species, mortality rates should show an increase during the usual adult life span, and there should be a fixed maximum life span with relatively little variation in the mean age of death in a protected environment. These conclusions follow from the principle that senescent effects will pile atop one another at a specific period late in life span, because, instead of predation and starvation causing a steady drop off in the number of individuals so that selection cannot operate, it is senescence itself that causes most mortality. There could be no selection for modifier genes that would push senescent effects beyond the age at which most individuals had died as a result of multiple senescent changes. In the same way that a sand dune is built from millions of grains that are carried to the top and then dropped where the peak blocks the force of the wind, the expressions of a multitude of discrete senescent effects are pushed by natural selection to the end of the life span where they collect atop one another because the force of natural selection is blocked by the sudden drop-off in population that is caused by senescence itself. This explains why the "Wonderful One-Hoss Shay" of Oliver Wendell Holmes:

> Went to pieces all at once, –
> All at once, and nothing first, –
> Just as bubbles do when they burst.   [17]

IMPLICATIONS

An evolutionary perspective on aging has important implications for basic attitudes and ideas about aging and death, for gerontology and research on aging, for ethics and social policy, and finally, for our more personal feelings about our own aging and death. The most important implications are those which have to do with our basic understanding of what aging is. Aging is not an accident. Aging is not an adaptation. Aging is not a disease, and it will not be cured. Medicine and changing social conditions have substantially extended the average life span, so many people hope, illogically, that it will extend the maximum life span. Relatively few people are aware that the maximal life span has not changed, at least in the last century ([10], pp. 72–77). Even fewer people realize that we will not be able to significantly extend the maximum life span. The effects that cause senescence are not only too numerous and separate from one another to be susceptible to much manipulation, but many of them also are part and parcel of what makes our bodies work. If they are disrupted, there are likely to be disadvantages in youth. Substantial disruption might well interfere with crucial parts of the body's machinery. An evolutionary perspective suggests that there is no clock that controls the rate of aging from some central point. Aging is the sum total of a multitude of changes. It *seems* coordinated, but this is explained by the action of natural selection, not by some central organizing mechanism. For once, the evolutionary biologists can warn their proximate biologist colleagues about teleological thinking. There is, in fact, no coordinated mechanism governing senescence; there are just many senescent effects that are expressed concordantly, because natural selection has pushed them together at the end of life.

There is no getting around it. Aging is here to stay. Aging is inevitable for individuals, not just in fact, but theoretically as well. Research on gerontology will not cure aging and is very unlikely to postpone it substantially. If we think that more money spent on aging research will accomplish these goals, we are fooling ourselves. Perhaps it is our wish to avoid these facts that has prevented consideration of aging in an evolutionary perspective. It is incredible that we have spent so many millions on senescence research without answering the basic scientific question of why there is senescence. Williams' theory has been available since 1957, but most doctors have never heard of it, and many gerontologists do not understand its significance. This may be a most interesting topic for a philosopher of science.

The next implications of an evolutionary view of aging are those that relate specifically to research and gerontology. As mentioned, almost all research has been focused on the proximate half of the problem. A small investment in evolutionary studies of senescence would pay big dividends in itself, and would also enhance our understanding of proximate mechanism. Proximate gerontologic research is also of great importance. It may not extend life, but it will improve the quality of life and may help us to find the cures for specific disease. Most of the diseases confronted by medicine today are diseases of senescence. There is every reason to believe that proximate research can lead to findings that will help specific individuals with specific diseases.

It is possible that some of the more common changes and diseases of aging may be pleiotropically linked to benefits earlier in life. Alzheimer's disease, atherosclerosis, and osteoporosis may be good candidates. Do people who have these diseases also have some advantages earlier in life? This could be the case despite numerous etiologic possibilities. For instance, even if Alzheimer's disease turns out to be caused by a virus, we might find that inability to resist the virus may be one effect of a gene that offers other benefits to the immune system. These examples are entirely hypothetical, but they illustrate the point.

Are there ethical implications of this view of aging? This is a delicate issue. Those who advocate an evolutionary perspective on human issues have often been criticized for directly drawing moral implications from biological facts. But many biological facts have no direct moral implications. To assume that they do shows the most primitive poor logic. Nonetheless, it seems to be a part of human nature for people to be tempted to take a biological fact and to conclude that what *is*, is what *should* be. They then use this precept as a guide for human choices. This is not only illogical, it is dangerous, because those who control a political system always seek justification for their favored position and their inordinate share of available resources. This kind of pseudo-biological rationale is surprisingly seductive for many people.

Even though biological facts are not independently sufficient to provide any ethical guidelines, an evolutionary perspective on senescence provides understanding that is essential for any discussion of ethics associated with aging. When combined with even a simple, ethical principle like "provide the greatest good for the greatest number," they do suggest possible changes in our behavior. In a large university hospital, the issues and contradictions are vivid. A 95-year-old terminally ill comatose woman may be put on a respirator for long enough to

wipe out the savings that might have offered her children new opportunities in life. On the other side, dialysis may not be recommended for
a 65-year-old person because of supposed advanced age. Many doctors
continue to treat death as the only enemy without knowing why there is
aging, or why death is inevitable. Understanding the evolutionary fact
of senescence and the inevitability of death might change attitudes and
behaviors. It tips the balance toward quality of a life, as against quantity
of life as a goal we should strive for. It suggests that physicians should
concentrate more on relief of problems that interfere with living, and
less on prolonging life. It does not suggest that the elderly should be
deprived of curative treatment when that is possible. It may have
implications for the increasing proportion of the Gross National Product
which is spent on health care and the increasing portion of this expenditure that is spent on attempts to prolong the lives of elderly patients,
even as the availability of other resources for the elderly are declining.

Finally, there are personal implications when we learn that there is an
evolutionary explanation for aging and for the inevitability of individual
death. Death turns out to be simply one move in the mindless but
perfectly efficient strategy of natural selection. Graffiti in the University
of Michigan Museum of Zoology succinctly summarize the central
point:

> Why are we born, only to suffer and die?
> Because those who suffered and died in the past,
> outreproduced those who didn't.   [2]

An evolutionary perspective on senescence offers new questions that
have important implications for gerontology, research, ethics, and our
personal understanding about aging and death. It is one example of the
use of evolutionary theory to better understand medicine and biology.
Darwin died 102 years ago, but the range and explanatory power of
evolution by natural selection is still not fully appreciated. Paradoxically, it appears to be precisely the issues of most crucial human concern
that have not been analyzed from an evolutionary perspective. Aging is
one example, many other problems are waiting. Each offers us an
opportunity to better understand our place in the natural world.

The University of Michigan
Ann Arbor, Michigan

BIBLIOGRAPHY

1. Alexander, R. D.: 1979, *Darwinism and Human Affairs*, University of Washington Press, Seattle, Wash.
2. Alexander, R. D.: 1981, 'Senescence: Explaining the Finiteness of Individual Existence', unpublished manuscript.
3. Bodmer, W. F. and Cavalli-Sforza, L. L.: 1976, *Genetics, Evolution and Man*, W. H. Freeman and Company, San Francisco, Calif.
4. Comfort, A.: 1979, *The Biology of Senescence*, 3rd ed., Elsevier, New York.
5. Clutton-Brock, T. H.; Guiness, F. E. and Albon, S. D.: 1979, *Red Deer: Behavior and Ecology of Two Sexes*, The University of Chicago Press, Chicago, Ill.
6. Cutler, R. G.: 1982, 'Longevity is Determined by Specific Genes: Testing the Hypothesis', in R. C. Adelman and G. S. Roth (eds.), *Testing the Theories of Aging*, CRC Press, Inc., Boca Raton, Fla., pp. 24–114.
7. Dawkins, R.: 1982, *The Extended Phenotype*, W. H. Freeman and Co., Oxford.
8. Dykhuizen, D.: 1974, 'Evolution of cell senescence, atherosclerosis and benign tumors', *Nature*, No. 5476, pp. 616–618.
9. Finch, C. E. and Hayflick, L.: 1977, *Handbook of the Biology of Aging*, Van Nostrand, Reinhold Co., New York, N.Y.
10. Fries, J. F. and Crapo, L. M.: 1981, *Vitality and Aging*, W. H. Freeman Co., San Frãncisco, Calif.
11. Gertzing, S. D.: 1957, 'Evidence of Aging in Natural Populations of Fishes', *Gerontologia* **1**, 287–305.
12. Haldane, J. B. S.: 1942, *New Paths in Genetics*, London.
13. Hamilton, W. D.: 1964, 'The Genetical Evolution of Social Behavior, Parts I and II', *Journal of Theoretical Biology* **7**, 1–52.
14. Hamilton, W. D.: 1966, 'The moulding of senescence by natural selection', *J. Theor. Biol.* **12**, 12–45.
15. Hayflick, L.: 1980, 'The Cell Biology of Human Aging', *Scientific American* **242**, 58–65.
16. Hinde, R. A.: 1982, *Ethology: Its Nature and Relations with Other Sciences*, Oxford Press, New York, N.Y.
17. Holmes, O. W.: 1857–1858, 'The Deacon's Masterpiece; or, The Wonderful 'One Hoss Shay', *The Autocrat at the Breakfast Table*, Boston, Mass.
18. Jackson, C. H. N.: 1940, 'The Analysis of a Tsetse-fly Population', *Annals of Eugenics (London)* **10**, 22–369.
19. Leopold, A. C.: 1961, 'Senescence in Plant Development', *Science* **134**, 1727–1732.
20. Mayr, E.: 1982, *The Growth of Biological Thought*, Belknap Press, Cambridge, Mass.
21. Medawar, P. B.: 1957, *The Uniqueness of the Individual*, Methuen and Co., Ltd., London.
22. Moment, G. B.: 1982, 'Theories of Aging: An Overview', in R. C. Adelman and G. S. Roth (eds.), *Testing the Theories of Aging*, CRC Press, Inc., Boca Raton, Fla., pp. 2–23.
23. Rose, M. and Charlesworth, B.: 1980, 'A Test of Evolutionary Theories of Senescence', *Nature* **287**, 141–142.
24. Sachar, G. A.: 1982, 'Evolutionary Theory in Gerontology', *Perspectives in Biology and Medicine* **25**, 339–353.

25. Sokal, R. R.: 1970. 'Senescence and Genetic Load: Evidence from Tribolium',
    Science **167**, 1733–1734.
26. Strehler, O. L. and Mildvan, A. S.: 1960, 'General Theory of Mortality and Aging',
    Science **132**, 14–21.
27. Tinbergen, N.: 1951, *The Study of Instinct*, Clarendon Press, Oxford.
28. Weisman, A.: 1881, 'The Duration of Life', in *Essays Upon Heredity and Kindred
    Biological Problems*, by A. Weisman, E. B. Poulton, S. Sehö, and A. E. Shipley
    (eds.), trans. A. E. Shipley and S. Schönland, Clarendon Press, Oxford, England, pp.
    1–66.
29. Williams, G.: 1957, 'Pleiotrophy. Natural Selection and the Evolution of Senescence',
    Evolution **11**, 398–411.
30. Williams, G. C.: 1966, *Adaptation and Natural Selection*, Princeton University Press,
    Princeton, N. J.
31. Wright, S.: 1932, 'The Roles of Mutation, Inbreeding, Cross-breeding, and Selection
    in Evolution', *Proceedings of the 6th International Congress of Genetics*, Ithaca, N.Y.,
    1, pp. 345–365.

SECTION II

PHILOSOPHICAL REFLECTIONS ON MEDICAL CARE PROVISION FOR THE AGED

DANIEL M. HAUSMAN

HEALTH CARE: EFFICIENCY AND EQUITY

In this essay I shall present and analyze the basic free-market approach
to the provision of health care. I shall not attempt to offer any definitive
judgment on its merits, for such a judgment, I will argue, depends both
on moral theory and on detailed empirical evidence, and I do not have
this evidence. Instead I shall first sketch out the general free-market
approach to the provision of medical care and indicate how the basic
approach must obviously be modified to deal with difficulties involved in
the provision of medical care, especially to the elderly. Then I will argue
that simple and global defenses or criticisms of the approach will not do.
Whether it would be a good thing to introduce more incentives into
health care provision is a difficult moral and empirical question. There
are no short cuts that provide good, quick answers.

The basic argument for leaving the provision of any good or service to
the market is simple and, in many of its applications, compelling. If
individuals have to shop for service 'S' and have to pay for 'S' themselves, then they will only purchase some of 'S' when they want it more
than other things they can purchase. Moreover, they will monitor how
much of 'S' is provided and how well it is provided, and they will avoid
sellers of shoddy services. Consumers will also, of course, shop around
and try to get the best value for their money. In those markets in which
consumers are able to judge the quality of the services they get and are
well informed about alternatives, providers of 'S' will have a powerful
incentive to economize and to provide better services than their competitors at the same price or equally good services at lower price. Otherwise they will lose their business to their competitors. Given certain
far-from-innocuous assumptions, it can be proven that competitive
markets are in a technical sense ideally efficient. In one interesting
sense, the result they bring about cannot be improved upon: there is no
way to make any individual better off without making some other worse
off.

If we substitute "medical care" for S, we have an argument for private
provision of medical care. We also have an argument against government provision of universal comprehensive health insurance. For if it

67

Stuart F. Spicker, Stanley R. Ingman, and Ian R. Lawson (eds.),
Ethical Dimensions of Geriatric Care, 67–78.
© *1987 by D. Reidel Publishing Company.*

costs individuals nothing (apart from time, anxiety, risks of humiliation, injury and even death) to seek medical care, they will not economize in doing so. Compared to the free-market benchmark, "too much" medical care will be sought, although this excess will be constrained by the aforementioned non-monetary inducements to avoid doctors and hospitals. More importantly, there will be no incentives on the part of health-care providers to provide the best value in medical care. Various incentives can, of course, be built into the comprehensive health insurance the government provides, but these are costly and blunt in their application.

So the argument goes that switching from our current messy system (with its similarities to a universal comprehensive health insurance system) to a free market system would increase efficiency. This conclusion rests on the theoretical argument concerning the nature of incentives in free markets and in insurance systems as well as on ample empirical evidence that considerable gains in efficiency are possible. Not only are some national health systems able to keep their population roughly as healthy as ours for a great deal less money, but within the United States one finds enormous diversity in health care practices (and costs) without systematic difference in outcomes.

But the provision of medical care has some peculiarities that make the application of the previous argument problematic. First, unlike purchases of clothing or cosmetics, expenditures on medical care are sometimes unavoidable, unanticipated, and large. Although individuals are sometimes responsible, at least in part, for their ailments, serious illnesses are nevertheless often like natural disasters: they strike and require enormous expenditures from individuals. These expenditures can be larger than individuals could possibly provide for by means of individual saving. And it would be foolish for individuals to attempt to provide for them by saving. Individuals are obviously much better off pooling the risks of serious illness rather than attempting to self-insure by heroic savings. So the rational economic man of orthodox economic theory will purchase health insurance.

And, one might think, given the logic of the free market, that there would be nothing inefficient about individuals doing so. But there is a problem or even a paradox here. In deciding to purchase insurance, an individual faces a relatively fixed set of prices for medical services and various insurance policies. Thus the net cost of medical insurance with a fairly small deductible is not much more than medical insurance with a large deductible or medical insurance that demands co-payments from the insured. If individuals are "risk averse" – that is, if they prefer not to gamble, they may then prefer to purchase relatively comprehensive medical insurance. But the effect of almost everybody's making a similar choice is greatly to increase the cost of medical services, since the incentives for providers and individuals to economize will be undercut. Efficient medical care is, I am suggesting, what economists call a "collective good," and, like other collective goods such as parks or water systems, it will not be secured by the self-interested actions of individuals. Another way of making the same point is to note that there are negative "externalities" in purchasing health insurance. In purchasing relatively comprehensive health insurance, one increases what others have to pay for health care.

So, quite apart from government regulation or interference, there is reason to expect that individuals will purchase health insurance policies that are so comprehensive that there will be few incentives for health care providers to control their costs. Supporters of market incentives in the provision of medical care might thus paradoxically (but without inconsistency) seek government intervention to prevent individuals from over-insuring.

A second complication which greatly aggravates the first is that we regard health care as a *need*. It used to be that individuals paid for their own health care when they could. When they could not do so, a hodge-podge of charitable arrangements provided for some minimal care. This system could not survive doctors' and hospitals' increasing medical capabilities and costs and the related moral recognition that competent health care was no longer a comparatively useless luxury to be made available only to those who could afford it, but instead a matter of life and death, a basic need. Something more than a voluntary insurance scheme was needed to supersede *ad hoc* charity and individual responsibility. The health care needs of those who were unable to purchase insurance or too foolish to do so are still needs.

It is not surprising that the solutions that were in fact adopted were those which most enriched doctors and hospitals. For not only do doctors and hospitals constitute a powerful lobby (which of course only grows more powerful as the health profession grows larger and richer), but this nation's official horror at "socialism" and its emphasis on individual responsibility provided mighty ideological weapons. Members of the health profession and their political allies argued that individuals, with the help and encouragement of both government and employers, should purchase comprehensive medical insurance. Government should intervene directly only to make sure that the medical bills of the indigent not go unpaid. Some government intervention is unavoidable, but it should be minimal.

It is obviously unacceptable to expect individuals to pay for their own medical care without even the help of insurance. There must be some way for individuals to pool the risks of serious illnesses, and government interference may be needed because permitting individuals to incur expenses privately without regulation may (as argued above) lead to great inefficiencies. Furthermore, medical care must be provided to individuals who are uninsured and unable to pay for their own care. The conclusion is particularly obvious with respect to the elderly, a large fraction of whom could not possibly afford adequate medical insurance. So no responsible commentator on the provision of medical care proposes eliminating insurance and assistance for the uninsured and placing the full responsibility for paying for health care directly on individuals.

Those who are impressed with the virtues of market solutions instead propose modifications in the diverse private and government insurance systems that have emerged during the last two generations and have so enriched the medical profession. Competition among insurance companies helps to some extent, but medical insurance underwriters are of necessity too large to monitor closely the performance of particular health care providers. The insurance provided by health maintenance organizations (HMOs) tends to be more efficient because these organizations can control the costs of providing medical care more directly, even though subscribers to such plans typically have no financial incentive to avoid seeking medical care. Apart from HMOs, deductibles and co-payments create financial incentives for individuals to avoid seeking medical care or to look for less expensive medical care providers. Some supporters of market solutions propose what is in effect a *much* higher deductible – perhaps as much as ten percent of an individual's income. Current proposals for a need-based criterion for Medicare are driven in part by the logic discussed here. From the perspective of the basic market model all of these proposals are going to be more efficient than comprehensive universal health insurance provided or sponsored by the government (with only a small deductible and little or no co-payment), for each proposal increases the cost accountability of health care providers.

As this argument correctly reveals, the comprehensive health insurance policies that many Americans possess provide few incentives for individuals to economize on their purchases of health care or for purveyors of health care to control the costs of the care they provide. Health care is consequently more expensive than it would otherwise be.

But this conclusion does not get us very far, for efficiency is not the only concern here, and even if it were, many relevant alternatives have been left out of the discussion. Proposed changes which are considerably less equitable than our current system of health care provision might be deemed unacceptable, regardless of their efficiency. For the sake of equity we may be willing to tolerate inefficiency. But even if one is mainly concerned with efficiency, the important question is not whether increasing incentives would lower costs – of course it would – but whether increasing market incentives is the *best way* to lower costs. For there are other alternatives that would also lower total costs for health care and increase efficiency. Two which worry supporters of market solutions are (1) comprehensive health insurance with cost controls and, worse yet, (2) socialized medicine. Should we attempt to revamp our current mix of private and public health insurance by creating further market incentives for individuals and institutions to economize and to control costs, or should we impose more government controls and management? Or should we do a bit of both? Or should we instead be paying more attention to factors such as diet, smoking, working conditions, or pollution, which are actual or potential causes of a large proportion of medical problems?

Those who defend the market incentive path use three main arguments to do so. First, they argue that incentives will be more efficient than government controls, since government controls require a costly bureaucracy that can be manipulated or corrupted by health care providers. Second, they argue that government controls must inevitably lead to unacceptable violations of individual freedom of choice. If it costs individuals nothing to go to the most expensive doctors and hospitals, some non-financial means of rationing these resources must be employed, and inevitably the government will wind up telling individuals which doctors they can see under which circumstances. Third, defenders of incentives argue that a market incentive system can be made fair. There are ways to avoid or to compensate for possible inequities that result from making individuals responsible for part of their health care costs. We might, for example, place a ceiling on how much of its income a family can be required to spend on health care each year, and we can provide subsidies to the indigent.

So there are basically three considerations offered in favor of an individualistic, incentives-based approach: (1) it is more efficient than government controls or management, (2) it is more consistent with the value we place on individual liberty, and (3) with proper constraints it can be made equitable. Let us examine each of these assertions in turn.

(1) (Efficiency) To assert that an individual-incentives approach is more efficient than alternative systems involving government controls or management is, at least in principle, to make an empirical claim. But it is not an empirical claim for which much direct evidence has been offered. There are instances in which one can make a strong case for the superior efficiency of markets, and to the extent to which health-care provision resembles these cases one has indirect evidence that a system of individual incentives will be more efficient than government controls. But even from the perspective of pure economic theory, there are reasons to hesitate about prescribing a market or quasi-market solution to health care funding. For the issue of externalities arises a second time in a significant way. My refusal to seek treatment may result in illness for many other individuals or in large later costs to me that as taxpayers they may be forced to share. On the other hand, vaccinating *my* child against mumps makes it less likely that *your* child will catch the disease. Such externalities are significant theoretically, since, as all economists would grant, markets may work badly in the presence of externalities. It may be that if individuals had to pay for their own medical expenses, they would go to the doctor too infrequently and would be worse off both physically and financially in consequence. I would like to see more adequate *evidence* for the claim that a market-incentive system really is more efficient than alternative arrangements that involve government controls or management.

When one is concerned with the provision of medical care to the elderly, the problems in assessing the efficiency of a market system as compared to a system of more direct government controls are greatly aggravated, for in either case the government is going to be greatly involved. Even if we modify Medicare to create more incentives for the aged to economize on their health care, we still have to provide Medicare to a large portion of the elderly, and we can demand little financial contribution from them. Without more government controls, very little can be done to lower the cost of the medical care of the aged. It might perhaps be argued that the greater efficiency of health care that would be brought about by a general incentive system would lower the cost of care to the aged as a by-product, even though many of the elderly would be covered by insurance policies resembling current ones. But given the large proportion of all medical care that goes to the elderly, the claim is unconvincing.

Notice that the main question is how best to get health care providers to control their costs. Financial incentives for individual patients to economize on their use of health care are of secondary importance, since there is a variety of non-monetary incentives that discourage people from visiting doctors or hospitals. I do not know how best to control health care costs, but it is not obvious that an individual-incentive system will be most efficient.

(2) (Liberty) In any case, narrow questions of efficiency cannot take one very far here, since ethical concerns weigh so heavily. The second argument is accordingly an ethical argument. Defenders of incentives maintain that our concern for freedom of choice should lead us away from a system of government control and management, since such controls will interfere with individual freedom. For example, Ellen may want to see super-specialist 'X' and may be willing to work nights to earn the money to pay the extremely high fee 'X' could command under an incentive system. But under a system of government controls, Ellen may simply not qualify to see 'X'. Are we willing to accept such a curtailment of individual freedoms or perhaps even such a violation of individual rights? And if we are not, is it not clear that we must follow the market incentive path?

I have two qualms about the argument. First, it exaggerates the extent to which government control or even management must interfere with individual choices. There is an enormous variety of ways in which government controls and management have been institutionalized, and under most of those which actually exist (let alone those which are yet to be invented), there are ways to make exceptions, so that Ellen, who is desperate to see 'X', gets to do so, even though she does not fall into the category of patients who would normally get to see 'X'. But, more importantly, the question is not whether government controls limit individual freedoms – obviously they do – but whether they do so *more* than a market-based system does. Even if (which may not be the case), under a market-incentive system, there is no doctor whom Ellen absolutely cannot see, there may be many doctors whom it would be a severe economic hardship or even a practical impossibility for Ellen to see. Will the average American have more real freedom to choose his or her health care under an individualistic-incentive system or under a system of more government controls? The constraints will be different, and defenders of the former are surely right that there will be significant constraints under a system of government controls; but what they need to show is that the constraints created by government control will be *more* severe or *more* objectionable from a moral point of view than are the financial constraints of a market-incentive system.

The only way to make such a case would be to rely on libertarian premises. Libertarians concede that granting people rights, or recognizing that they have rights. imposes obligations on others. But the only obligations libertarians recognize are obligations *not* to interfere. Ellen's right to life imposes on others the obligation not to kill Ellen, but it does not impose on others any obligation to feed Ellen if Ellen is starving or to succor Ellen if Ellen is ill. No doubt (except in the view of extremists such as Ayn Rand) 'X' is not a very charitable or admirable person if 'X' lets Ellen die when 'X' might easily have kept Ellen alive, but in doing so 'X' does not violate Ellen's rights: Ellen has no cause for moral complaint. If one can accept such a stark view, then there is no place for rights to health care, and there is a clear moral asymmetry between the actions of a bureaucracy that limits Ellen's right to purchase health care from whomever she chooses and the impersonal operations of a market that may at least as effectively prevent Ellen from getting the care she wants. But this is (fortunately) one area where libertarianism conflicts with most people's moral intuitions. There is no such clear moral asymmetry: if someone is allowed to die of appendicitis simply because he or she has no money, we have not only a failure of charity, but an injustice as well. In other words: individuals have rights to some health care, and a market system that in a completely impersonal way prevents them from getting that health care can violate their rights no less than a government bureaucracy can.

(3) (Equity) Although particular proposals to introduce more financial incentives may be extremely unjust, I do not believe that there is any sound argument showing that every possible incentive system must be inequitable. The only general objection on the grounds of equity that can be made to all incentive systems is that the rich will be able to take advantage of more expensive health care than will the middle classes or the poor. Some of the greater expense of the health care the rich

purchase may have nothing to do with its quality. But unless the rich are fools, we can assume that they will obtain better health care than the poor. Some might question this assumption since overtreatment can be at least as bad as undertreatment, but these doubts seem to me unpersuasive. At least for the sake of argument, let us suppose that the rich get better health care than the poor. Is there anything objectionable about this?

I think that in principle there is. In my view our recognition of health care as a basic need demands as an ethical ideal not only that poverty should not prevent anyone from receiving competent health care, but also that riches and poverty should be completely irrelevant to the quality of the health care one receives. We should, I believe, think of health care as similar to basic liberties, such as freedom of religion, of which everyone should have an equal share regardless of economic status, rather than as similar to housing or food, of which everyone should only be guaranteed some minimum amount. For health care differs critically from food and housing and other goods and services. In visiting a doctor, just as in visiting a restaurant or a car dealer, one is purchasing something, and one has a choice of how much to spend. But nobody thinks that the difference between chopped liver and *paté de foie gras* or the difference between a Chevrolet and a Cadillac and, indeed, no car at all, is a matter of life and death. Food and some freedom of movement are needed, but *paté* and Cadillacs are not. But the services of the best surgeon one can find for one's suffering child are arguably not a luxury, but a need. It may be that such needs cannot all be met. Someone may have to be denied needed health care. But wealth should be no more relevant in deciding who must be denied health care than in deciding how to ration food in a famine. Since health care consumption is driven so largely by basic human need, it is unfair (even if it turns out to be politically unavoidable) for the rich to have better health care than the indigent.

Yet even from an egalitarian perspective such as mine, there is no ground to condemn incentive systems in general. For, first of all, they may themselves be egalitarian in their operation if the distribution of resources among citizens is itself egalitarian. But, more importantly, it is just not politically feasible (no matter how desirable from an ethical perspective) to prevent the wealthy from getting at least what they deem to be better health care. The most one can reasonably aspire to do is to offer everyone some *decent minimum* level of competent care. Even a purportedly egalitarian system like the British National Health System only offers roughly equal care to those not rich enough to opt out of it. Anything more egalitarian is utopian. So there is no compelling objection, in general, to all incentive systems on grounds that they will permit the rich to purchase better health care than the poor.

In reaching this conclusion, I am, of course, not passing judgment on the equity of any particular proposal to increase economic incentives in the provision of health care. Many of these proposals are terribly unfair and would badly harm the poor. Those who recognize health care as a basic need should strenuously oppose such proposals. But a proposed incentive system must be judged independently on its own merits. There is no basis for condemning all such systems as inequitable.

One further complication deserves some discussion: Many people are, to varying degrees, responsible for their medical problems. People smoke, and they eat and drink too much or too much of the wrong substances. They take dangerous drugs, refuse to use seatbelts, jump out of airplanes, and choose or are coerced into dangerous occupations. If individuals had to bear the costs of their unhealthy life styles, either directly or through much higher insurance premiums, then they would have an incentive to take better care of themselves. Furthermore, it seems only fair that those who take risks with their health (or with the health of their employees) should have to pay the resulting costs.

The problem is, however, that our liberal traditions and institutions make it infeasible in most cases to impose the medical costs of unhealthy life styles on those who create them. Insurance companies cannot, for example, find out whether individuals fasten their seatbelts. Without much more intrusion into the private lives of individuals than we are collectively willing to tolerate, it is impossible to monitor life styles sufficiently to insure that those who live dangerously pay the costs of doing so. To discourage unhealthy life styles requires non-market interventions or intrusions into our private lives. The best liberal option is education.

The ethical questions concerning the provision of health care are likely to become more pressing as doctors continue to learn how to do more for their patients. One unfortunate consequence of the many splendid recent improvements in medical knowledge is that applying this knowledge often requires very expensive training, equipment and facilities. Of course there is no way to predict the course of technological progress, and it may happen that technological means will be found to make health care much cheaper. But the trend is toward more expensive treatments. In the case of geriatric care, this trend is exaggerated by the increasing shifting of the burdens of long-term care from the individual family to hospitals, nursing homes, and other paid caregivers.

As this trend continues, the problems in deciding how to provide health care efficiently and fairly will become more difficult. Notice that the various proposals for introducing more market incentives into the provision of health care are incapable of controlling the astronomical costs of major procedures and long-term care, since such costs will surpass the ceiling on what all but the richest individuals can be expected to pay. There will be scarcely more incentive for cost control in such cases than there is now. Unless we surrender our concerns with equity, the only way that such costs will be contained will be through government controls. An individual incentive system cannot suffice. We are going to need government control and, to some extent, management as well.

But the ethical problems of the increasing efficacy and expense of innovations in health care seem to be the really serious problems. Perhaps we will be fortunate and the cost of medical technology will not keep increasing; it is arguable that much of the current expense is due to its imprecision. But if we are not lucky, we are inevitably going to have to make ghastly choices that will determine who will *not* receive needed treatment. Either impersonally through a much crueller market system than is currently defended or more directly through government mandate, we collectively are going to have to decide that some people cannot have expensive treatments that could have saved their lives.

In such circumstances the essence of the problem of health care provision lies naked: What is at issue is a collective moral decision concerning how much of our resources *should* go to health care and how that health care *should* be distributed to individuals. To lay too much weight on questions of efficiency or to attempt to hide behind the market would not only be unjust, but it would betray moral cowardice. Let us hope that the moral problems of health care provision never become too painful – that we are lucky enough that the medical profession can continue to learn how to do more to keep us healthy without that knowledge becoming too costly to benefit everybody. But whatever mix of market incentives and government controls we settle on, we must not shy away from our collective moral responsibility to decide what kind of medical care system we want to have.

Carnegie-Mellon University
Pittsburgh, Pennsylvania

ACKNOWLEDGMENT

I am grateful to Ernie Alleva, Catherine Kautsky, John Kautsky, Jonathan Moreno, Jonathan Pressler, David Rosner, Stuart Spicker and to members of the audience who heard my original presentation for their helpful criticisms and suggestions.

BARUCH BRODY

WHOLEHEARTED AND HALFHEARTED CARE: NATIONAL POLICIES *vs.* INDIVIDUAL CHOICE

This whole symposium, as its title indicates, is taking place at a time in which our societal sense of what is appropriate medical care is being challenged by our economic concerns about how much health care our society can afford. For America, this is a new experience, one which makes all of us very uncomfortable. I hope in this paper to first raise the level of discomfort, and then to make certain suggestions that are the beginning of an attempt to try to lower that level.

I. THE ECONOMIC BACKGROUND TO THE CURRENT DISCUSSION

The simple facts are that America spent $322 billion on health care in 1982, an amount equal to 10.5% of the Gross National Product ([7], p. 1). This is a dramatic increase from the $93.5 billion spent on health care ten years earlier, in 1972, an amount equal to 7.9% of the Gross National Product (GNP), and an even more dramatic increase when compared to the $41.7 billion spent on health care in 1965 (a year whose significance we will return to), a mere 6% of that year's Gross National Product ([7], p. 4).

The response to these simple facts is a national outcry. Medicare, we are told, is going broke, and the elderly will not be able to receive the health care they require. Medicaid is causing a major crisis for both the federal government and the states which jointly pay for it. Health insurance costs, rising in response to health care costs, are causing, we are told, a major problem for American industry which must compete in the international marketplace. Finally, some health care is paid for by recipients, and they are confronting a crisis in bearing their share. The conclusion drawn is that America must stop these trends; health care costs must be contained. The only questions are how to do so and at a cost to whom.

The first thing that we need to do is to get a clearer picture as to what has happened. Consider, to begin with, the period of 1971–1981 ([6], p. 10). Around 60% of the increased costs in that period were due to

79

Stuart F. Spicker, Stanley R. Ingman, and Ian R. Lawson (eds.),
Ethical Dimensions of Geriatric Care, 79–93.
© *1987 by D. Reidel Publishing Company.*

general inflation and to population growth. The remaining 40% were due to factors specific to the health care industry. What are they? Three deserve to be mentioned: (a) medical care price increases in excess of general price inflation; (b) more visits to physicians or hospitalizations per capita; (c) more services per visit or per hospitalization. This last category is, of course, a residual category, and is the largest category. Moreover, some of (a) is related to (c) as well, since some of those excess price increases were due to higher labor costs due to greater use of labor, and this may reflect more services. So (c), normally calculated as 21% of the increase for hospital visits, for example, is probably closer to 25% of the increase for hospital visits.

Are factors (b) and (c) desirable or not? Consider the following argument: the Medicare and Medicaid systems became firmly established in the late 1960s. By the early 1970s, people were really using them to get the health care which they traditionally would not have received. That's what (b) represents, and we should welcome it. Moreover, increasingly sophisticated medical practices in hospitals have provided more good health care for people being hospitalized ([7], p. 4). That's what (c) represents, and we should welcome it as well.

This argument can be put in perspective if we consider the broader period of 1950–1980. The first half of that period saw the first major growth in the percentage of GNP (4.4% in 1950 to 6% in 1965 – a 1/3 increase) devoted to health care. That period marked the spread of hospitalization insurance, which made hospital care available to so many, and the first great spurt of the new medical technologies which gave doctors so much more to offer to patients. The period of 1965–1980 (6% of GNP to 9.5% of GNP – a more than 50% increase) continued both of those trends, with the passage of Medicare and Medicaid in 1965 and the ever-increasing growth of medical technology, and added to those trends the problem of rapid inflation. These two major trends, more people served and more services, are to be welcomed. So why the sense of national crisis?

Saying all of this is not, by the way, arguing for the claim that there is no waste in our current health care expenditures. Everyone has his own horror story to tell about unnecessary procedures performed for the economic benefit of the provider, defensive medicine practiced at great cost in the vain hope of avoiding malpractice suits, sloppy hospital practices that encourage wasteful expenditures, etc. But all of the stories, and even the studies which back them up, do not produce a

WHOLEHEARTED AND HALFHEARTED CARE 81

sense of national crisis. It is the gross expenditure figures which do, and we have just seen that their interpretation requires careful analysis for they may primarily reflect, once adjusted for inflation, desirable trends. So again, why the sense of crisis?

I think that there are two ways of answering this question, and that both point to realities that need to be joined to form a fuller picture. The first, and more cynical, reflects upon the pressures that these historical trends are placing on major decision makers in our society, politicians and business leaders. The second, somewhat less cynical but ultimately more threatening, reflects upon the future implications of current trends.

Suppose it is true that most of the growth in health care expenditures, once these figures are adjusted for inflation, represents more Americans getting more good medical care. Still, someone has to pay for it. The parties who have immediately absorbed most of the increased costs are the state and federal governments who supply one form of third party payment and industries who pay for another form. In 1929, patient direct payments were 88.4% of all health care expenditures. In 1950, they were still 65.5%. By 1965, with the spread of private health insurance, they had fallen to 51.8% (private health insurance, usually funded in large part of employers, had picked up that entire change). In 1982, they were only 31.5% (with public expenditures picking up that entire change) ([7], p. 4). So we need to look at the pressures on government officials and on employers. Neither is the primary recipient of the benefits of better health care for more Americans. Yet the former must find tax revenues to pay for those benefits, and voting for increased tax revenues is dangerous to the political health of elected officials. The latter must raise prices or lower profits to cover increased fringe benefits costs, and, at least in an international setting, the former is dangerous, while the latter is undesirable. So, runs the cynical argument, those who are most central to decision-making are facing a crisis, and the perception of a crisis is promoted by them as a way of getting the public to accept structural changes (often undesirable from a health care perspective) which will limit costs.

As I said above, I believe that there is some truth to this cynical account, much as there is some truth to the claim that there is waste in the health care system which needs to be abolished. There is, however, still a third part to the story of the current crisis, and that is the most threatening part of the story. Its basic claim is this: most of the inflation adjusted growth is due to good goals, and not to waste. But those very goals, providing better health care to more Americans, have not been fully satisfied yet, and there are reasons for believing that fully satisfying them in the future will be harder and harder. If we continue with our current attempts to meet those goals, however desirable it may be to do so, we will wind up with legitimate health care expenditures consuming so much of our GNP that we will not be able to meet other legitimate social needs. That is why we are facing a crisis in the provision of health care in America.

This point, which will be the basis of the rest of our discussion, deserves elaboration. There are several facts which make it likely that the pursuit of the goal of providing all reasonably beneficial health care to all Americans who need it will involve great future costs: (a) we still have about 10% of the population who are uninsured and at great risk if they get ill; (b) while the elderly and the poor are getting much more medical care, we would fool ourselves if we thought they were getting all the care from which they would benefit; (c) we will have many more elderly Americans, and many more very elderly Americans, in the future, and all of these people will require more health care since people in those age brackets make the greatest demands on the system, particularly for long-term care; (d) we have every reason to expect further medical breakthroughs which will offer more improved health care, but often at greater costs, especially to the aged who need this care. All of these factors, leaving inflation aside, will mean that health care costs will consume still a greater percentage of our GNP. The crisis then becomes one of choosing what we will not have if we continue to increase the percentage of our GNP devoted to health care.

It is true that a recent estimate [6] by analysts at HCFA projects that only 12% of our GNP in 1990 will be devoted to health care, which is not such a large growth from the 10.5% of 1982. This projection is, however, based upon two major economic assumptions. One is that inflation will remain relatively lower in the 1982–1990 period than it was in the 1972–1982 period, and the other is that real gross national product will increase far more rapidly in the 1982–1990 period. The former will help keep health care costs down, while the latter will provide more funds to pay for health care. This helps explain why we are projected to spend only 12% of GNP in 1990 on health care, even though projected health care costs for 1990 are $690 billion. If either or both of these economic assumptions are false, then that percentage of the GNP will be much worse than 12%. So I would not rush to be comforted by that projection.

I conclude that some of our current crisis in health care financing is a function of wasteful expenditures that could be resolved by more careful control of expenditures. Some of the sense of crisis has been, I have suggested, created by those who are reluctant to make the decisions required to finance further growth in health care expenditures. But much of the crisis reflects a concern about how far we can go in the future to meet real medical needs without shutting off other needs. It is this last portion of the health care crisis, particularly as it impacts on the elderly, which I will be addressing in the rest of this paper.

II. CURRENT METHODS OF CONTROLLING COSTS AND THEIR ULTIMATE FAILURE

Recent discussions ([9], pp. 378–449) of ways to cut health care expenditures have centered around the debate between the regulatory approach and the market approach. The regulatory approach attempts to control health care expenditures by prohibiting certain behavior thought to be costly and inappropriate (e.g., building new facilities unless a need for them can be demonstrated) and mandating other behavior thought likely to moderate costs (e.g., mandating peer review of decisions to perform certain procedures, to insure that the procedure is actually required). The market approach attempts to control health care expenditures by providing economic incentives which encourage cost-saving behavior. These incentives may be directed to providers (e.g., limiting the reimbursement they will receive) or to recipients (e.g., requiring them to pay for more of their health care).

We shall not enter at this point into the wide-ranging current controversy as to which of these approaches is preferable. There are empirical issues involved concerning which is more likely to succeed. There are value questions involved concerning the merits or demerits of the market, and the value of freedom and the merits of social order. I am concerned here with making a different point about both of these approaches.

These approaches seem most appropriate when the enemy being confronted is waste in health care expenditures. Specific forms of wasteful behavior can be prohibited by regulation. Other forms can be controlled by mandated control mechanisms designed to stop the wasteful while allowing the medically appropriate to move forward. There remain the questions of whether regulations work and at what cost to other values (e.g., freedom) we hold, but there is something appropriate about regulations to avoid wasteful expenditures. A similar point can be made about market mechanisms. If financial incentives, either for providers or for recipients, will lead them to avoid wasteful behavior, then implementing such incentives seems perfectly appropriate. There remain the questions as to whether such incentives actually work, and at what cost to other values (e.g., fairness to poor recipients), but there is something appropriate about using financial incentives to avoid wasteful expenditures.

The situation starts looking very different when we are no longer confronting waste but rather increased expenditures required to provide needed, but often expensive, health care to all those who need it, even if they cannot afford to pay for it on their own. Even a well-regulated and well-structured health care system, which effectively prohibited many forms of wasteful health care expenditures, and which provided effective incentives which discouraged other forms of wasteful health care expenditures, will, we have argued above, face a crisis in health care expenditures in the years to come. Such a crisis, if it is to be met by traditional means, calls for more than the traditional regulations and the traditional market mechanisms. Let me elaborate.

Consider, for example, the traditional peer-review program. It was designed to insure that certain unnecessary procedures were not carried out simply for the economic benefit of the providers, and it did that by mandating a review of the decision to perform the procedure by others who had the expertise to evaluate the need for the procedure but who did not gain from its being performed. Clearly, such a mandated program was designed to avoid waste. Suppose, now, that you wanted to use such a program to deal not with waste but with legitimate needs. You might, for example, limit the number of procedures of a certain type to be performed each year, and authorize a peer-review board to oversee the decisions as to who gets the needed procedure and who does not. This would be a very different matter. The criteria would no longer be whether the procedure was needed, but that plus other factors. What would they be? Who would have the expertise to make those decisions? I do not say that these questions cannot be answered. All that I am saying is that using regulations to meet a problem not rooted in wasteful expenditures is entering into a very different type of game.

Consider, as another example, the newly emerging prospective reimbursement scheme. It was clearly designed to discourage providers from providing unnecessary services by not reimbursing them for these extra services. It does this by reimbursing providers with a flat fee, adjusted each year for inflation, for treating a patient with a particular diagnosis. The thought clearly was that providers would no longer perform unnecessary services since they would have to absorb the cost of providing that unnecessary service. Suppose, now, that you wanted to use such a program to deal not with waste but with legitimate needs. You might, for example, make no adjustment, or at least an inadequate adjustment, for new and expensive but better approaches for treating a patient with a particular diagnosis. This would be a very different matter. We would now need to decide how much medical improvement we were willing to pay for, and we would need special criteria for making this decision. I do not say that such decisions cannot be made. All that I am saying is that using financial incentives to meet a problem not rooted in wasteful expenditures is entering into a very different type of game.

I have not, of course, used these two examples arbitrarily. We are seeing the emergence of a new form of peer-review with mandated goals (not yet quotas) and the emergence of severe limitations on the adjustment to the Medicare prospective reimbursement scheme. These are no longer attempts to avoid wasteful increases in health care expenditures. These may be the first indications of America's entering into a national rationing policy, a policy designed to limit the growth in health care expenditures by refusing to provide certain services to certain recipients even though they would benefit from them.

This is the crucial point and we need to keep it in mind. As our population ages, and as our abilities to provide good but expensive medicine increase, we will be facing a crisis in the growth of health care expenditures that neither traditional regulations nor traditional market mechanisms were designed to confront. They were designed to provide alternative acceptable ways of eliminating wasteful expenditures. It is easy to form a social consensus against waste. The only question we have to face in fighting waste is the mixture of regulation and market mechanisms that best does the job and satisfies our other values. It is much harder to form a social consensus about what forms of useful health care should be denied to which recipients, and it is therefore difficult to see how we can use either traditional approach in dealing with the real emerging crisis in the growth of health care expenditures.

All of this helps us explain what is happening in the organ-transplant area [2]. Liver transplants and heart transplants are now joining kidney transplants as the treatment of choice (best and quite efficacious) for certain conditions. But these are expensive forms of treatment. For a while, the traditional third parties avoided assuming their cost by claiming that they were still experimental procedures. That is no longer a viable option. There is no social consensus for rationing such procedures, so we are rapidly seeing the emergence of third-party responsibility for paying for such procedures. Neither traditional regulations nor traditional market mechanisms could prevent this development. And would we want them to? Why shouldn't those whose lives can be saved by these techniques live on with their help? Isn't paying to allow these people to live many good years part (perhaps the most important part) of our commitment to the sanctity of human life? But then, how can we control health care expenditures once we get past the control of waste?

I promised that the first part of this paper would raise the level of discomfort. I believe that this promise has been fulfilled. I hope in the remainder of this paper to fulfill my second promise, to make suggestions that are the beginning of an attempt to try to lower that level of discomfort.

III. RATIONING: NATIONAL POLICY vs. INDIVIDUAL CHOICE

The following are the main claims that have emerged from the first part of our analysis: (1) America's commitment to providing to all of its citizens the health care which would benefit them may be the major cause of the non-inflation related increase in health care expenditures; (2) This is not something to regret, even if it poses delicate problems for major decision makers; (3) Demographic trends about aging and the explosion in medical capacities are likely, in the future, to challenge that commitment because we have other social commitments; (4) All of this will probably lead us to refuse some efficacious medical care to some recipients. Call this rationing (as opposed to eliminating wasteful expenditures). The question now becomes: How shall we ration? Who shall decide on what basis that certain potential recipients shall not receive certain forms of potentially efficacious health care, that certain potential recipients shall receive halfhearted as opposed to wholehearted life-prolonging medical care?

There are three approaches which we might adopt to this type of decision making: (a) A national medical policy limiting certain types of care to certain recipients; (b) A national budgetary policy which would limit the total amount of care that could be provided but which would leave actual decisions not to provide care to individual providers (hospitals or physicians); (c) A national social policy encouraging more patient input into the choices to treat or not to treat and encouraging more attractive alternatives to wholehearted traditional medical care. I want to argue for (c) and against (a) and (b).

Aaron and Schwartz, in their recent study [1] of rationing of health care in England, have shown the extent to which the British rely upon a type (b) policy. Consider, for example, the fact that the British do not use TPN (total parenteral nutrition) as much as we do. On a population-corrected basis, they°spend less than one-fourth as much as the U.S. does ([1], pp. 52–56). There is no national policy limiting in any way the use of TPN, and since its use involves no special capital facilities or major equipment, its use is not restricted by any national policy restrictions on capital expenditures. It's just that each hospital's pharmacy has a budget, and when the use of TPN rises, the chief pharmacist is able to effectively stop further use of TPN on the grounds that it blocks other care. In one large teaching hospital, Aaron and Schwartz report, the chief pharmacist's pressure led the staff to agree that a maximum of six adult patients could receive TPN at any one time. The relevant consultants (specialists) then had to decide who would get that form of care and who would not, even if they could benefit from it. Similar patterns emerge in many other British approaches to rationing, even in the much-discussed kidney dialysis case.

What is wrong with such an approach? It seems to me that there are many difficulties with it. To begin with, it works in England, to the extent that it does, because English patients are very passive and accept, either out of ignorance or out of respect for the physician's authority, the physician's claim that nothing more can be done. Often, those who do not accept this passively, and insist on further care, get it. Such an approach is unlikely to work in America, where patients are increasingly knowledgeable and non-passive and where the threat of malpractice suits provides them with a powerful weapon. Secondly, and this is a related point, the whole thrust of American medical ethics and medical jurisprudence in recent years has been to encourage patient involvement rather than patient passivity. Those who accept this as a positive development can hardly encourage a form of rationing that depends on passivity. Thirdly, the English system allows for too much arbitrariness on the part of individual decision-makers and too much variation between them. In effect, it allows senior physicians the right to make fundamental value choices subject only to the constraints of their budget and the need to avoid too much public outcry. Fourthly, it works only in a system in which there is a central national health care budget. Even in England, you can avoid rationing by going into the private sector, which is not subject to the central budget. It is hard to see how such a system could be implemented in America, given the wide variety of third party payers.

Consider now the idea of a national medical policy which would limit certain types of care to certain patients, a policy which would say that some should not receive wholehearted care even though they might benefit from it. This would at least avoid the arbitrariness and excessive variability possible in the British system. We could even see how it might be implemented in America as an agreement between all public third party providers. Private third party providers might have those limitations built into their standard policies, but individual recipients might purchase (without any tax benefits from doing so) policies that extended their coverage to these extra treatments. But there would still be many problems with such an approach: (a) In practice and in theory, it would still depend on patient passivity in accepting rationing choices. We are unlikely to have that passivity in America, and we don't want to encourage it; (b) It would depend on an ability to form a social consensus about who should be denied what forms of health care, and it is highly unlikely that such a consensus could emerge (more on that point below); (c) The most vulnerable groups in our society, the elderly and the poor, are likely to be hurt the most by such policies. In particular, given that the elderly need so much health care, it is likely that rationing will be applied in ways that most directly harm them. This might seem acceptable to these who view old age as a disease, or who adopt the view that the elderly have had their chance and must now move over to make way for the younger, but it is not acceptable to all those who would stress the value of human life and the real contributions that the elderly can make to a society.

I turn then to policy (c). Policy (c) is designed to involve patients rather than depend on their passive acceptance. It involves no central health care budget, so it is applicable in America. It avoids arbitrariness and unfair variations by leaving decisions in the hands of those whose care is at stake. Finally, it lessens the fear of preying on the vulnerable. So it offers, I submit, the best hope for a way of rationing care. But I need to say something more about what policy (c) involves.

The advocates [3] of policy (c) begin with the observation that there are many patients currently receiving expensive health care that does extend their life, and is in that sense efficacious, but for whom the extension of life is a questionable benefit just because the resulting quality is so low. These include at least those dying painfully and without dignity from a slow-acting terminal illness, those whose death is imminent and who are being kept alive in an artificial environment for a short period of time, and those whose mental status has deteriorated to a very low level but who can live for a considerable period of time as long as their recurring infections and other problems are treated (the persistent vegetative patient is the clearest example, but some of the advanced dementia patients fall into this category). The advocates of (c) go on to observe that the provision of expensive therapeutic or life-prolonging care to these patients is often an automatic choice, and the patients in question (or those who speak for them) are given little if any choice about this care. They suggest that many of those patients (or those who speak for them) would reject such forms of care if they were given the chance and were given the option of wholehearted palliative care. They conclude that a policy promoting this might be our best bet for an appropriate cost-saving form of rationing.

Before we go any further, I need to say something about whole-hearted and halfhearted care. The choice to be presented to these patients is not a choice between wholehearted life-prolonging care or halfhearted life-prolonging care. The latter sounds terrible, and is worse in practice. It needs to be a choice between wholehearted life-prolonging care and wholehearted symptom-relieving and dignity-preserving care. So the choice that advocates of policy (c) would present to the patients in question is a choice between two forms of whole-hearted care.

Would this type of rationing be enough to alleviate the future cost pressures on our health care system? That would depend on the answer to three questions: (a) how many patients fall now (and in the future) into these categories; (b) how many of them (or those who speak for them) would choose palliative care; (c) how much this would save per patient now and in the future. Advocates of policy (c) believe that many patients now fall into these categories, and the very trends that pose the future crisis indicate that there will be many more in these categories in the future. They also believe that many will choose palliative care and that the savings are great and will be greater in the future as the cost of life-prolonging care grows. But these are just beliefs. Studies need to be done to show that they are true.

One final point. I have argued elsewhere [4] that the ideal scheme for the public support of health care would be a general adequate redistributive scheme that made no special provisions for health care other than providing the indigent with sufficient funds so that they could, if they wished, purchase, among other things, some level of health care in advance. This scheme, which I call a quasi-libertarian approach, is based on an emphasis on individual choice, interfering with that only at the level of the initial redistribution of funds. That is its great merit. I do not expect to see it adopted, since it goes against so many established interests. I do wish to note, however, that its spirit is present in the current proposed scheme (c) for rationing life-prolonging health care, a scheme that rests heavily on freedom of choice.

IV. PROBLEMS

Nobody who advocates any solution to a major crisis should be taken seriously if he suggests that there are no problems with his proposed solution. I do wish to be taken seriously. Therefore, I will end this paper with a list of possible problems that my proposed solution might encounter.

I see the following as the major potential difficulties: (a) Health care providers might not be comfortable with providing patients and/or families with the options about which we are talking – our proposal presupposes that, at certain crucial points where the patient's condition justifies raising the question of whether continued aggressive life-prolonging measures are justified, health care providers will raise these questions and discuss them with patients and/or their families. Patients and/or their families might sometimes be able to take the initiative, but they usually will not, either because they do not have the knowledge that enables them to recognize the situation or because it is emotionally difficult for them to come to grips with the situation. So we will need to count on health care providers' taking that initiative. Will they? There is no doubt that many health care providers currently do not [5]. In part, this reflects the awkwardness of having such a conversation (telling someone or his family how close he is to dying and how little hope there is left). In part, this reflects the feeling that one doesn't want to stop treating patients (and the presupposition that that's what you are doing when you stop trying to prolong their life). In part, this reflects a feeling of defeat, and who wants to admit that? And, I suspect, it also reflects an unwillingness to accept the reality of death, with its implication that everyone, including the health care provider, must eventually die. This, then, is our first major problem. In light of all of the reasons leading health care providers to hesitate about raising the alternative of palliative care, can we count on them to do so on a regular enough basis to enable our proposal to have the impact it could potentially have on health care costs? (b) Health care recipients and/or their families might not be willing to accept the alternative of palliative care – our proposal presupposes that, were the choice given to them, patients and/or their families would usually accept the choice of palliative care. This presupposition fits in nicely with a certain picture very prevalent in the literature of medical ethics and medical jurisprudence. It is a picture that derives from a set of classic cases (e.g., Quinlan), and it portrays health care providers as eager to continue heroic life-prolonging measures long after there is any point to them and patients and/or their families struggling to prevent this. This portrayal is sometimes accurate, but it often is not. In many cases, it is the patient (and even more so, the family) who insists on trying any possibility long after the physicians have given up [5]. Why? In part, this reflects an unwillingness to accept the reality and inevitability of imminent death or of a prolonged low quality of life for oneself of for someone whom one loves. In part, this reflects a fear that the patient will be deserted to die in pain and indignity if the physicians stop trying to prolong the patient's life (a fear that the "patient will be put on the back burner"). In part, this reflects all sorts of family anger about the condition of the patient, guilt about their past relations with the patient, etc. I have seen the parents of a persistent vegetative patient, angry about what fate had decreed for their only child, continue to demand aggressive life-prolonging care five years after the patient's entering that state, insisting that "a miracle is always possible." They are not all that unique, just a little more extreme than most. This, then, is our second major problem. In light of all of the reasons that lead patients and/or their families to continue to demand life-prolonging care long after it can promise much of value, can we count on them to accept the alternative of palliative care on a regular

enough basis to enable our proposal to have the impact it could potentially have on health care costs? (c) We might not even be able to have a
social consensus about the types of cases in which we wish to encourage
presenting patients and/or their families with a choice of life-prolonging
care *vs.* palliative care – our proposal presupposes that we can identify a
range of cases in which it would be appropriate to withhold life-
prolonging care if the patient and/or his family concurs and in which it is
appropriate to raise that question with the patient and/or his family. I
used to be confident that we could count on a consensus about this.
After all, this would not be a consensus about withholding life-
prolonging care, just a consensus about when to offer that option to the
patient and/or his family. I am now less confident. Consider, after all,
what happened in Minnesota when this issue was raised by the Task
Force on Supportive Care [10]. They were proposing, in effect, a policy
like ours, of mere palliative care if that was what the patient and/or his
family wished, and a policy like ours, of raising that question in appropriate cases with the patient and/or his family. They identified three
classes of patients (terminally ill and imminently dying, severe and
irreversible mental disability, and severe and irreversible physical disability) whose condition justified providing the patient and/or his family
with such a choice. Immediately, the Nursing Home Action Group [8]
challenged their proposal, insisting that the *choice* of mere supportive or
palliative care should be provided only for patients whose death is
imminent from an irreversible terminal illness. We are beginning to see,
at the end of life, the same quarrels that the "Baby-Doe" regulations
have revealed at the beginning of life. So perhaps the consensus we are
seeking, and which our proposal presupposes, may not exist about
enough cases to allow our proposal to have the impact it could potentially have on health care costs.

   These are the major potential difficulties with our proposal which I
can identify. Our proposal can work only if society can agree about a
significant enough set of cases in which providers should offer recipients
a choice about life-prolonging versus palliative care, only if providers
will regularly offer that choice in these cases, and only if recipients
and/or their families will often accept the choice of palliative care. Each
of our objections (a)–(c) challenges one of those necessary conditions.
(c) claims that there is no consensus about a significant enough set of
cases, (a) claims that providers won't often enough offer the choice even
if there is a consensus, and (b) claims that recipients won't often enough
accept the choice even if it is offered.

Objections are not necessarily refutations. This is certainly true about these objections. They raise difficulties, rather than refutations. To some degree, it's an open empirical question as to the extent to which these difficulties stand in the way of our proposals even now. It is an even more open question as to whether better education about the capacities and limitations of health care, education directed to the public (to providers and to recipients) couldn't make our proposals even more workable. What I want to end with is really a challenge: given the unacceptable alternatives to our proposal, and given the reality of the health care funding crisis which we will be facing, shouldn't we attempt to find out, by some pilot studies, whether a national policy of promoting individual choices about life-prolonging medicine isn't a viable policy for appropriate health care in America's near future?

Center for Ethics, Medicine
  and Public Issues
Baylor College of Medicine
Houston, Texas

BIBLIOGRAPHY

1. Aaron, H. and Schwartz, W.: 1984, *The Painful Prescription: Rationing Hospital Care*, Brookings, Washington, D.C.
2. Annas, G.: 1985, 'Regulating the Introduction of Heart and Liver Transplantation', *American Journal of Public Health* **75**, 93–95.
3. Bayer, R. *et al.*: 1983, 'The Care of Terminally Ill: Morality and Economics', *New England Journal of Medicine* **309**, 1490–94.
4. Brody, B.: 1981, 'Health Care for the Haves and Have-Nots', in E. Shelp (ed.), *Justice and Health Care*, Reidel, Dordrecht, pp. 151–59.
5. Evans, A. and Brody, B.: 1985, 'The Do-Not-Resuscitate Order in Teaching Hospitals', forthcoming in *Journal of the American Medical Association*.
6. Freeland, M. *et al.*: 1983, 'National Health Expenditure Growth in the 1980's', *Health Care Financing Review* **5**, 1–31.
7. Gibson, R. *et al.*: 1983, 'National Health Expenditures, 1982', *Health Care Financing Review* **5**, 1–31.
8. Hoyt, J. and Davies, J.: 1984, 'A Response to the Task Force on Supportive Care', *Law Medicine and Health Care* **12**, 103–05.
9. Starr, P.: 1982, *The Social Transformation of American Medicine*, Basic Books, New York City.
10. The Task Force on Supportive Care: 1984, 'The Supportive Care Plan – Its Meaning and Application', *Law Medicine and Health Care* **12**, 97–102.

THOMAS HALPER

COMMENTARY ON BARUCH BRODY'S ESSAY

Perhaps because it confirms humanity's general hope for progress, the impact of medicine on society has for decades been a popular theme, both in official rhetoric and private speech. Of course, not all health gains have been due to medicine, but the public, its officials, and the media have tended to confuse the two, and to credit increased life expectancy and improved vigor to the health care establishment. Further, medicine, unlike science in general, is usually thought of as an unalloyed good. Where science's name is mottled by nuclear weapons and environmental depredations, medicine's is barely blemished at all.

In the process of acquiring this aura, medicine has helped to shape the individual's outlook on himself, his society, his very future. Indeed, so essential is contemporary medicine to the notion of modernity itself that it is quite impossible even to imagine a society advanced in all areas save that. Finally, medicine has consumed such vast quantities of resources, financial, intellectual, and technological, that its very appetite has had major effects on the life of the nation. America, after all, spends half again as much on health care as on that legendary devourer of money, national defense.

In the tangle of upbeat clichés, however, it is easy to lose sight of the fact that society has had an enormous impact on medicine, too. For it is society, through its political and bureaucratic agents, that has determined that medicine merits the immense allocations to which it has grown accustomed, and it is society that encourages medicine to move in this direction rather more aggressively than in that.

It is one of the central virtues of Professor Brody's paper that it so thoroughly appreciates the interactions, subtle and not so subtle, between medicine and society. Indeed, the futility of trying to understand one without taking into account its relations with the other seems to me one of its key underlying themes, and it is a theme of considerable power and impact. All this is evident in virtually every paragraph, and most tellingly, in my view, in a brief but exceedingly cogent discussion of the problems inherent in applying the British approach to rationing to the United States.

95

Stuart F. Spicker, Stanley R. Ingman, and Ian R. Lawson (eds.),
Ethical Dimensions of Geriatric Care, 95–103.
© 1987 *by D. Reidel Publishing Company.*

More generally, I think that even a casual reader would find this paper an ambitious, candid, and immensely stimulating piece of work. It not only confronts the question of why health costs in America have zoomed; it has the audacity to propose a solution to the problem. Meanwhile, value issues that arise naturally along the way are not evaded but receive pithy, often unconventional answers, and defects in his proposed solution are noted with a frankness and force that are as impressive as they are disarming. Most importantly, perhaps, from the outset the reader is provoked into a re-examination of previously held attitudes and beliefs that continues long after the actual reading is finished. It is hard to imagine a person interested in health care who could not benefit from studying this paper.

The paper itself, I believe, can usefully be divided into two parts. The first is concisely summarized in four key points.

It is difficult to disagree with *point number one*: "America's commitment to providing to all of its citizens the health care which would benefit them may be the major cause of the non-inflation related increase in health care expenditures." The other points, however, seem to me to raise rather stickier questions. Let me try to discuss them in turn.

Point number two: "This [aforesaid commitment] is not something to regret, even if it poses delicate problems for major decision makers." This, it is earlier explained, is because Medicare and Medicaid have permitted more visits to physicians and more hospitalizations per capita and because "increasingly sophisticated medical practices in hospitals have provided more good health care for people being hospitalized."

Is this to be regretted? No one, of course, could regret the goal of a healthy citizenry. This is not to say, however, that one can usefully examine benefits in isolation from their costs. When over a quarter of the entire Medicare budget goes to maintain patients in their last year of life, most of that in their last month, it is not difficult to imagine wiser alternative expenditures of the funds. Indeed, Professor Brody himself appears implicitly to acknowledge this in his own trenchant discussion of life-prolonging care. The simple assertion of Medicare's providing "more good health care," then, conceals as it reveals, and hardly seems an adequate description, given the size and complexity of the program. But this, I am afraid, appears to me part of a more general refusal of the paper to confront and take seriously the issue of cost. Thus, third party payers are described as governments and industries, as if taxpayers, employees, and consumers did not really bear the burden. A recognition that people have many other goods and services they wish to purchase in addition or even in preference to health care is not given sufficient prominence. This latter issue seems to me particularly salient, raising as it does such topics as paternalism, inter-temporality, pluralism, public goods, and coercion. I am certain that Professor Brody has many fascinating things to say about these topics, and I hope that he will forgive me for wishing that he had said them.

Nor does applauding the goal of a healthy citizenry preclude regret over the choice of means employed to achieve that end. It is possible to argue, for example, that America should not commit itself to provide all of its citizens health care (the entitlement principle), but only those who are unable to provide it for themselves (the need principle). In the first place, the need principle is cheaper to apply. For however need is defined – and the endless arguments on line drawing would surely have enormous ethical and practical consequences – it must be less inclusive than society-wide entitlement. It may also, arguably, be more just to redistribute resources in the interest of the less well off than simply in the interest of the sick, the injured, or the aged. For example, given that the elderly as a class significantly exceed the national average in net income and total assets – not to mention the receiving of payments in kind, special benefits and discounts, "off the book" earnings and gifts, and so on – the arguments for an age-based transfer system like Medicare will not strike all observers as equitable [6]. Medicare, after all, pays almost as much to the elderly earning $32 000 per year and over as to those earning $10 000 and under ([15], p. 205).

It is even possible to argue that health services simply should not be offered to the poor *qua* poor, who instead should be given money – quarrels over amounts would certainly be fierce – with which they could purchase health services or anything else they chose. Such an approach would recognize that the problem of the poor is lack of money and not lack of health (or other) services, and avoid the paternalistic assumption that the poor cannot be trusted adequately to protect their own interests. It would also discourage the development and maintenance of a vast governmental and non-profit industry, whose manifest goal of helping the poor has tended to be displaced by the latent goal of helping themselves.

It is possible to argue, too, that a national commitment to good health need not entail assigning health care such a rapidly growing share of the gross national product. After all, good health is not principally a product of the efforts of the medical establishment, but historically has also involved sanitation, education, housing, traffic and industrial safety, preventive health measures, and so on ([11], [13], [14]). Increasing the tax on cigarettes by a dollar a pack, for instance, might contribute more to the health of the citizenry than doubling the number of coronary artery bypass surgeries, and the development of the child-proof medical bottle cap doubtless has saved more lives than the perfection of the liver transplant can ever promise. Put differently, despite the hundreds of billions of dollars consumed by Medicare, it is worth noting that death rates from all causes other than heart disease and stroke were almost the same in 1980 as in 1965 – and even the changes in heart disease and stroke may have been due as much to non-medical factors (improved diet, smoking, and exercise patterns) as to medical ones (better hypertension drugs, open heart surgery, etc.). And Berkman and Breslow [1] reported that a Health Practices Index (including cigarette and alcohol consumption, physical activity, etc.) that omitted health care entirely was a good predictor of individual health status and mortality. All this suggests that some of the funds currently being spent on health care might better advance health if diverted to other purposes, like education, traffic safety, pollution reduction, and so on.

Nor is it evident that the principal problems posed by this national commitment to health care will be those posed for decision makers. On the contrary. What seems plain is that elected officials and bureaucrats are normally rewarded for granting benefits and often punished for denying them. This is because the relatively small number of beneficiaries feel far more intensely about the issue than the much larger number of their fellow citizens who must pay the bill. The temptation to provide an ever increasing array of benefits, as a consequence, has proven irresistible, for while the beneficiary's return may be large, the taxpayer's share is invariably very small. As these small shares accumulate as one program is added to another, however, the total tax burden may become rather heavy, as the taxpayer comes to resemble a leaf that a multitude of insects is devouring, silently and happily, tiny bite by tiny bite. As Everett Dirksen used to say, "A billion dollars here, a billion dollars there, and pretty soon you're talking about real money." Politicians and bureaucrats may not be "the primary recipients of the benefits of better health care for more Americans", but they are certainly major secondary recipients, and it has never been easy to persuade them to trade the pleasures of saying "Yes" for the risks of saying "No".

Point number three: "Demographic trends about aging and the explosion in medical capacities are likely, in the future, to challenge that commitment [to providing health care for all] because we have other social commitments." The key element here is the projected growth in the proportion of the elderly and the very elderly, the kinds of people who "make the greatest demands on the system, particularly for long-term care."

Whether the aged will so overburden the system, however, is not self-evident, either. Of course, the proportion of elderly will increase, but a much better predictor of medical cost is not years from birth but years to death. A recent study of Medicare enrollees, for example, revealed that the average reimbursement for patients in their last year of life was 6.6 times (and in their next-to-last year 2.3 times) the rate of those who survived at least two years [12]. Fuchs has shown, however, that the percentage of persons over 65 who will die within five years has remained virtually constant since 1965 [3], suggesting that the widely held cost projections on the coming burden of the elderly may seriously overestimate the problem. There thus may be less force to Professor Brody's point than is immediately apparent.

Point number four: "All of this will probably lead us to refuse some efficacious care to some recipients," a practice Professor Brody terms "rationing." The regulatory and market approaches may control waste, he observes, but only rationing can address the larger issue of scarcity.

Here, too, I believe that questions arise. One is: what is meant by "waste"? Double billing surely is waste. Performing ultrasound tests to get pictures for baby albums is waste. Using CAT scanners to evaluate minor headaches is waste. But is dialyzing a marginally demented elderly patient waste? And what of keeping a young patient in an intensive care unit after brain surgery, not because he is critically ill but because he is at risk of becoming critically ill? The risk of internal cranial bleeding may be small, but without the prompt, appropriate response of the skilled ICU staff, the bleeding might also be fatal. Is spending over $1000 per day for this care waste? And what of expenditures on treatment or research that seemed reasonable and prudent but that failed? Should every failure be written off as waste, leaving human perfection our standard? And is the cost of reducing waste always worth the effort? For a serious attack on waste can hardly avoid compromising physician/scientist autonomy, stifling delays and paperwork, and heightening fear of risks and preoccupation with bureaucratic procedures that may smother the spirit of inquiry and innovation.

The answers are not always obvious, for waste must refer not merely to abuse and incompetence but also to differences over priority rankings, which in turn pivot on value judgments that may be highly debatable. My "waste" may be your "need."

Waste, therefore, is not only often subjectively defined but also commonly tied to scarcity. After all, if there were unlimited resources, there could be no waste. But if the term "waste" implies a preference for cost effectiveness, such an approach has never won a wide following in America among either health care providers or consumers. Physicians, that is, are neither taught nor expected to offer benefits only so long as they exceed costs; on the contrary, the traditional ethic has required them to offer benefits and not focus on costs at all. The growth of third-party payments has reinforced this ethic to the point that physicians have become so free from the fear that high prices will drive away patients that, at least in hospitals, they appear quite uninformed as to the charges for the tests and treatments they order ([2], [9], [10]). Cost effectiveness, in any case, is easier to advocate than to implement. The ordering of tests and the prescribing of drugs, after all, is not such an exact science that the physician can always be certain as to what is needed and what is not, and as tests and drugs multiply, this sense of uncertainty may well grow with them. Moreover, the physician surely knows that if a vital test or drug was omitted for its apparent lack of cost effectiveness, this explanation will hardly satisfy his patient, his family, or his malpractice attorney. In some societies, like the United Kingdom ([4], [5]), cost effectiveness is well established within the health services; in the United States, it is not.

Relatedly, unless "waste" is construed as misallocation and not simply abuse and incompetence, it seems to me simply mistaken to assert that regulatory and market approaches "seem most appropriate when the enemy being confronted is waste." For issues of allocation – at the macro and micro levels – have been addressed by regulatory and market approaches with varying degrees of success since time out of mind. What is the new policy of establishing standard payment rates for Medicare patients based on the diagnoses of each patient but a regulation designed to induce hospitals to control costs due to abuse, incompetence, *and* inappropriate priority rankings – and can anyone imagine that DRGs will represent the last effort in this direction?

More basically, point number four may well be merely a truism, for the very existence of scarcity would seem to entail some sort of rationing. Not all medical care is available to everyone – a rural hamlet in Arkansas cannot offer the resources of a Boston or New York – and not everyone is sufficiently knowledgeable and aggressive (or has sufficiently knowledgeable and aggressive family and friends) to take advantage of the resources that are available. We may bewail this, but inasmuch as scarcity probably inheres in the human condition, we can hardly treat it as a recent or transitory phenomenon. Rationing, in short, is not new; only its increasing visibility lends it that illusion.

The second part of Professor Brody's paper deals with approaches for implementing the inescapable rationing. Here, he argues for "a national social policy encouraging more patient input into the choices to treat or not to treat and encouraging more attractive alternatives to wholehearted traditional medical care." After a particularly searching and concise discussion of the British practice, he notes that many patients receive expensive medical care that leaves them with a low quality of life – so low that these patients or their families would surely prefer less expensive palliative care. Then, with characteristic honesty, he confronts three major problems raised by this approach.

Although these objections are stated with impressive clarity and vigor, I believe that further troubles may also exist. For one thing, the discussion seems to me to neglect the costs of patient-centered decision making. Medical decisions are frequently highly technical matters, and expanding the role of actors with limited pertinent knowledge, skills, and experience, not to say a possibly blinding vested interest in survival, escape from pain, and so on, may not be the best path to appropriate decisions. When even experts may find themselves guessing about the cost-effectiveness of new therapies and technologies, how can we expect patients to produce soundly based judgments?

In addition to the costs to society and medical personnel, moreover, there are the costs to the patient himself. Consider the stress of participating in a decision as to whether one lives or dies, stress that itself may well have medical implications. Consider a patient, desperate, depleting his meager energy reserves, pleading for his life, but, not knowing what information may help him, feeling compelled to bare his soul and

become an object of pity. Consider a patient thus bereft of privacy and
self-respect, scarred even in victory, unendurably embittered in defeat.
The role of paternalism in medicine is a complex one [7] and the case for
a patient-centered approach may need more elaboration than it receives
here.

For another thing, there is the question of uncertainty. Terminal care,
which Professor Brody advocates and toward which I am generally quite
sympathetic [8], proceeds on the presumption of certain, imminent
death. Indeed, administratively, it has been defined as less than three
months of projected survival. Palliative care, on the other hand, envi-
sions the possibility of much longer survival, perhaps measured in years,
and even entertains the chance of an occasional cure. For many patients, it is intellectually a simple matter to assign them to the terminal
category, but for many others, it is not. And where there is uncertainty,
it may seem quite rational to the patient, his family, and his physician to
opt for palliation. It is not, then, merely an unwillingness to accept
defeat or death itself that may lead a patient to prefer palliation over
terminality, but also a rational fear that terminal care might prematurely
and irreversibly dismiss an opportunity for a meaningfully prolonged
life. Of course, significant levels of uncertainty are not always present,
but they are present often enough to forestall the mechanical invocation
of psychology to explain why patients may shrink from terminal care.

Furthermore, it is not clear that a patient-centered approach would
generate sufficient dollar savings to carry the entire rationing burden.
Professor Brody concedes this, almost in passing, but never provides
data to reinforce his position. The reader, therefore, is left feeling that
the argument has come perilously close to the promise of a free lunch;
that is, it seems to be maintained that sufficient funds can be raised at
the negligible social cost – perhaps, it is even a social benefit – of
choosing palliation over cure in appropriate cases. This seems to me to
be too good to be true, and as my father used to say, when something
seems too good to be true, it is probably not true.

As I review my remarks, I fear that I may have left the impression of
unremitting disagreement. But a commentator who can find nothing to
complain of is an embarrassment to the profession; worse yet, he has
too little to say. None of these remarks, though, should deflect the
prospective reader from a fascinating and utterly absorbing paper. I am
greatly pleased to have been given the opportunity to discuss it and, by
so doing, I hope, to have added a little to its visibility.

BIBLIOGRAPHY

1. Berkman, L. F. and Breslow: 1983, *Health and Ways of Living*, Oxford University Press, New York.
2. Dresnick, S. J.; Roth, W. I.; Linn, B. S.; Pratt, T. C.; Blum, A.: 1980, 'The Physician's Role in the Cost-Containment Program', *Journal of the American Medical Association* **241**, 1606–9.
3. Fuchs, V. R.: 1984, 'Though Much Is Taken: Reflections on Aging, Health, and Medical Care', *Milbank Memorial Fund Quarterly* **62**, 143–66.
4. Halper, T.: 1985, 'Life and Death in a Welfare State', *Milbank Memorial Fund Quarterly* **63** (Winter), 52–93.
5. Halper, T.: 1985, 'End-Stage Renal Failure and the Aged in the United Kingdom', *International Journal of Technology Assessment in Health Care* **1** (Jan.), 41–52.
6. Halper, T.: 1984, 'Aging Policy in the Eighties: Second Thoughts on a Strategy That Has Worked', in S. F. Spicker and S. Ingman (eds.), *Vitalizing Long-Term Care*, Springer Publishing Co., New York, pp. 3–13.
7. Halper, T.: 1980, 'The Double Edged Sword', *Milbank Memorial Fund Quarterly* **58**, 339–426.
8. Halper, T.: 1979, 'On Death, Dying, and Terminality: Today, Yesterday, and Tomorrow', *Journal of Health Politics, Policy and Laws* **4**, 11–29.
9. Kelly, S. P.: 1978, 'Physicians' Knowledge of Hospital Costs', *Journal of Family Practice* **6**, 171–72.
10. Kirkland, L. R.: 1979, 'The Physician and Cost Containment', *Journal of the American Medical Association* **242**, 1032.
11. Levine, S.; Feldman, J. J.; Elinson, J.: 1983, 'Does Medical Care Do Any Good?' in D. Mechanic (ed.), *Handbook of Health, Health Care, and the Health Professions*, Free Press, New York, pp. 394–404.
12. Lubitz, J. and Priboda, R.: 1982, *Use and Costs of Medicare Services in the Last Years of Life*, Office of Research and Office of Statistics and Data Management, Health Care Financing Administration, Baltimore, Md. (mimeograph).
13. McKeown, T.: 1976, *The Role of Medicine: Dream, Mirage or Nemesis*, Nuffield Provincial Hospitals Trust, London, England.
14. McKinlay, J. B. and McKinlay, S. M.: 1977, 'The Questionable Contribution of Medical Measures to the Decline of Mortality in the U.S. in the Twentieth Century', *Milbank Memorial Fund Quarterly* **55**, 405–28.
15. Wilansky, G. R.: 1982, 'Government and the Financing of Health Care', *American Economic Review* **72**, 202–7.

TEO FORCHT DAGI

REVIVAL, RESUSCITATION, AND RESURRECTION: THE RIGHTS OF PASSAGE

This essay is about decisions concerning the resuscitation of the aged, how they mirror general social policy towards the elderly, and how they characterize and occasionally define tacit and otherwise unrecognized attitudes and prejudices. It examines whether resuscitation should be carried out automatically whenever death threatens, or whether, in certain situations, the obligation to resuscitate is suspended. It asks whether there is only a *prima facie*, rather than an absolute obligation to resuscitate; whether the attainment of a certain age modifies this obligation by restricting either the obligation to resuscitate or the right to be resuscitated; and whether there is any change when death threatens unexpectedly rather than after a long and painful illness. Finally, it questions whether the patient's or the family's wishes or beliefs enable or compel the modification of this obligation in any way, and whether there is a distinct obligation to respect, elicit, or even to leave advance directives (e.g., a "living will"). There are no simple answers to any of these questions.

There is a *prima facie* obligation to resuscitate that constitutes both a tenet of medicine and a universal moral imperative [70]. For this reason, no modification of this obligation can be contemplated without restructuring a full hierarchy of social values. Thus, when the Neasden Memorandum in Britain tried specifically to discourage the resuscitation of patients over the age of 65, it encountered tremendous hostility, in part because the care of the aged is universally acknowledged as a charitable act [43]. Paradoxically, while public indignation prompted the rapid disavowal of that particular policy, similar rules have continued to exert widespread influence on professional conduct and the elderly are often denied equal access to medical care.

Two important concerns are frequently raised when the resuscitation of the aged is discussed: first, that the resuscitation and continued support of the ailing aged would strain the limited resources of the health care system by creating a population of non-contributing survivors; second, that resuscitation will be performed uncritically on

105

Stuart F. Spicker, Stanley R. Ingman, and Ian R. Lawson (eds.),
Ethical Dimensions of Geriatric Care, 105–126.
© 1987 *by D. Reidel Publishing Company.*

patients who would prefer to be allowed to die, simply because technical advances have enabled the detection and preservation of increasingly subtle signs of residual life. A plea for "death with dignity" often follows, embodying an expression of the ancient quest for control of one's fate by appointing the time and manner of one's death, as well as the more obvious challenge to the wisdom of routine resuscitation.

Baer expands the plea for "death with dignity" by suggesting that physicians have a distinct obligation to protect the elderly from unwanted resuscitation. "It is my belief," he writes, discussing the treatment of patients over 65 with obtundation and dominant hemisphere strokes, "that if the type of stroke patient we are considering here was aware that sustaining his biological existence by intravenous fluid and antibiotics throughout the usual week or even ten days of poststroke coma could easily result in a nursing home existence rather than true rehabilitation, he would, if he could speak, say, 'Let me have good nursing care – and nothing else'. . . . I personally would rather trust to 'survival of the fit' to survive rather than submit myself to the fallible dictates of scientific medicine" ([6], p. 381). Baer concludes that death can be a fitting end to life, rather than an enemy always to be vanquished. Similarly, Negovsky differentiates between "untimely death, when the human organism's vital functions have not yet been fully exhausted" and the obligation to resuscitate is all-powerful, and "the natural termination of life in old age," when the obligation to resuscitate is quite weak [61].

There is still another side to this question. Although the aged may be sometimes subjected to undesired resuscitation, they are at least as vulnerable to altruistically motivated but inappropriate decisions *not* to resuscitate. While *biologic* age may predispose to illness and disease, *chronologic* age can predispose to well-intentioned medical neglect ([4], [54], [67], [72]). Restricting the resuscitation of the elderly is neither an isolated nor a simple decision: when considered in detail, the issues surrounding such restrictions attain a significance quite disproportionate to their frequency, and their costs. This essay will focus on the problems inherent to decisions limiting the resuscitation of the aged, and the implications of such decisions for social policy on a wider scale.

REVIVAL, RESUSCITATION, RESURRECTION

I. THE SCOPE OF THE PROBLEM

A. Contradictory Attitudes to Aging

Why are there some apparent contradictions inherent in our attitudes towards the aged? We are committed to improve the medical care of the elderly, yet we worry about the economic consequences of an aging population. We strive for longevity but, because of the discordance between health and longevity, we fear the ravages of senescence.

It was fairly uncommon to attain old age until the past century. The threat of illness was constant and the abilities of medicine were limited: only the healthy could aspire to live more than four or five decades. The primary mission of medicine, therefore, was to increase the average life expectancy by forestalling death; quality of life issues were secondary. This task could be accomplished in three ways: disease prevention, disease treatment, and resuscitation. Although resuscitation, the most dramatic of the three, was the last to be reliably effective, success in the other two gradually increased the size and the age of the patient population to whom resuscitation was at least potentially applicable. Longevity, however, has not become synonymous with health. We have neither inherited nor developed a satisfactory mechanism for confronting the partial (and some say Pyrrhic) victory of achieving senescence while eliminating neither the disease nor the bodily degeneration of age ([19], [20], [21], [23], [27], [35], [42], [50], [65], [86], [87]).

B. The Meaning of Cardiopulmonary Resuscitation (CPR)

Does resuscitation have the same meaning in every context? Safar argues that 'reanimation,' 'revival,' 'resuscitation,' and most similar terms can be used interchangeably [71]. I disagree.

One commonly accepted definition of resuscitation is "the ability to rescue people from the brink of death by restoring life – beating heartbeat and breathing, . . . the revival of a living being from apparent death" ([63], p. 232). Here the use of 'resuscitation' to describe efforts to restore life *after* death seems excluded. Although the moment of death can be estimated with fair precision in retrospect ([34], [35]), one can only speak of *impending* or *imminent* death before the fact: for the *exact* moment of death may not be precisely determinable even as it occurs. A person on the brink of death, therefore, must be considered alive, and so must a person who is only "apparently" dead.

Elsewhere, in contrast, resuscitation is used *specifically* to refer to efforts that commence after apparent death: "CPR is quite literally a technique for bringing the dead back to life. It is initiated when what is recognized in the popular mind as death – the cessation of pulse and respiration – has already occurred. It snatches the patient from death's grasp and, when successful, may restore him to life within minutes" [29]. I believe that CPR *before* death and CPR *after* death are different entities and carry different moral obligations linked to the meaning of death, and to traditional imperatives to prove the reality of death.

C. Revival, Resuscitation, and Resurrection

'Resuscitation' describes efforts to save a person from the brink of death. At this stage he is considered alive. Resuscitation can also be *attempted* after a person has died. It is preferable to speak of 'resuscitation' when a patient is restored to life from the brink of death, but to speak of 'resurrection' when a patient recovers *after* death. By definition, death is irreversible, so that no success can be anticipated. Nonetheless both the effort and the outcome are so frequently discussed that they need a distinct designation. 'Resurrection,' because of its religious implications, may not be the best word, but I know of no better term, and 'resurrection' has achieved some measure of acceptance in the medical literature. Hartikainen, for example, chose 'resurrection' to describe a patient who was declared dead following prolonged but unsuccessful CPR. Twenty minutes later, on his way to the morgue, the patient began to breathe spontaneously and recovered [31]. 'Resuscitation' seems inadequate to describe this state of affairs (although it could be argued that the diagnosis of death in this patient was simply incorrect).

There are several situations in which resurrection may be almost routinely attempted, though not identified as such. When an unsterile thoracotomy is performed in the emergency room, for example, to close a gaping bullet wound in the heart of a pulseless and apneic patient, the patient may be dead by all accepted criteria, yet certain operational or moral considerations in the provision of emergency services may preclude the declaration of death until a last rescue has been attempted.

'Resurrection' is also useful to describe the effort applied to a person who is declared dead, is apparently dead, but is in fact alive. After a recent plane crash, the ranking fire surgeon was physically restrained by police from treating a person who showed signs of life because the victim had already been pronounced dead. Death disenfranchises: it is generally assumed that death will only be declared *after* it has occurred. Even a spurious declaration is binding. The obligation to resuscitate becomes suspended, and efforts that earlier would have been called 'resuscitation' must be renamed.

The terms 'resuscitation' and 'resurrection' can be distinguished operationally. In resuscitation, the victim is assumed to be only *apparently* dead; in resurrection, the victim is adjudged to be *truly* (irreversibly) dead. There are strong indications that resuscitation which is unsuccessful after 30 minutes or more is futile [9]. Any effort lasting longer than 30 minutes, therefore, "drifts" into attempted resurrection.

The clinical literature points to a category of resuscitative efforts in which initially satisfactory cardiac output and respiration serve as a positive prognostic variable ([9], [21], [42], [51], [52], [62], [81]). Emergency assistance is required for what is essentially a self-limited condition. The victim is clearly alive, and recovers self-sufficiency during the course of treatment. I propose that this category be called 'revival' to distinguish it from resuscitation and resurrection. 'Revival' covers a variety of clearly delineated clinical situations (e.g., a faint, or an airway obstruction while eating) in which the threat of death is distant, vital signs are stable (even prior to intervention), and full recovery is expected.

D. Death, Declaration of Death, and Apparent Death

'Revival,' 'resuscitation,' and 'resurrection' are inextricably linked to the meaning of death. The common law defines death as the *irreversible* cessation of all vital signs. The judgement of when death has taken place is generally left to the physician. Statutory criteria may differ from the common law definitions in specific jurisdictions. Currently, two general sets of criteria coexist: cardiovascular criteria, under which death may be declared when blood pressure, pulse, heartbeat, and respiration cease to be detected; and brain death criteria, justifying the declaration of death when irreversible cessation of *all* cerebral function is confirmed. There is an inescapable ambiguity inherent to the simultaneous validity of both criteria ([7], [8], [11], [32]).

Death and the declaration of death are not identical. The declaration of death is a legal affidavit stating that death has occurred. It can be false, mistaken, or mistimed. It has no necessary relationship to the *biological* phenomenon of death.

Apparent death refers both to conditions that are very difficult to differentiate from death, and to situations in which death is suspected, but unproven. Apparent death and apparent life are two sides of the same coin. Both allude to a possible disparity between the appearance of the victim and reality, and both declare that the apparent phenomenon of death is open to proof.

E. CPR as a Test of Death

CPR can act as a test of death. Bedell, for example, argues that death after cardiac arrest cannot and should not be declared unless and until resuscitation has failed [10]. Unfortunately, there is no sure way to decide when resuscitation has failed [48]. The most widely cited criteria do no better than recommending that efforts at resuscitation be abandoned after "findings of cardiovascular unresponsiveness," when resuscitation has been carried out in a "timely and appropriate manner so that these efforts represent an adequate test of the responsiveness of the victim's cardiovascular system" [73]. It would be extremely helpful to define a population in which CPR would *always* fail: in this group, there would be no obligation to *test* for death any further (chronologic age has not proven a valid criterion). Is there an obligation to test for death when death is not unwelcome? This is one of the many pragmatic questions regarding the formal and procedural relationships of resuscitation to the definition of death that remains to be considered ([9], [27], [45], [58], [79]).

II. THE OBLIGATION TO RESUSCITATE

We have seen how the complex notion of 'resuscitation' can be analyzed into three categories with different operational meanings and how resuscitation in the generic sense can be used as a test of death. We turn now to the origins of the obligation to resuscitate in order to determine whether and when this obligation can be justifiably modified by biologic or chronologic age.

A. Historical Roots

Historically, the obligation to resuscitate arose from an imperative to prove that death had occurred, from a fear of premature interment after a malicious or mistaken misdiagnosis of death, and from a desire to rescue those who only seemed to be dead. This imperative was reinforced by ancient beliefs in "latent life" during which the soul could be recalled, and by the observation that such attempts were occasionally successful ([3], [5], [36]; [66], pp. 135, 512–513; [69], pp. 265–268; [80]). Clerics were presumed to have special abilities to produce and reverse apparent death in theurgic societies ([73], Act IV, i:93–106). Alchemists and physicians established similar claims in other settings.

Historical accounts suggest that some early attempts at resuscitation were astonishingly effective. John Hunter, for example, empirically advocated the establishment of waterside "receiving houses" for the resuscitation of drowned sailors [36]. In 18th-century England, it was common knowledge that some corpses began to breathe while carted to the cemetery in rude, unsprung wagons. Relatives of executed criminals would hasten to the gallows to cut down the body and consult the nearest "hanging surgeon" in the hope of restoring life [68]. A powerful obligation to test death became a part of common folk wisdom, constituting, perhaps, one of the earliest examples in the resuscitation literature – and in medicine as a whole – of the ethical aphorism "can implies ought."

By the end of the 19th century, attempts at intubation and cardiac massage were commonplace ([18], [57]). All told, a great deal of energy was invested in discovering how to restore the dead to life; how much of this effort arose from aspirations to immortality, how much from an interest in advancing scientific medicine, and how much from an imperative to "prove" death and recover those who were only *apparently* dead cannot be accurately reconstructed. To maintain perspective, however, it is useful to remember that the ability to successfully and consistently restore heartbeat, blood pressure, and respiration is less than 40 years old ([43], [44]; [63], pp. 232 ff.).

The *historical* obligation to attempt a rescue from death generally required that death be an *error* of fate, unbidden and unfair. Sufficient signs of residual life had to be present to warrant the attempt. The *modern* obligation is not fundamentally different, and an unending fascination with the "near death" experience bears witness to the primitive fears that continue to exert their influence in modern times ([58], [61], [79]). The modern obligation has been expanded, however, to include universal competence in CPR and adequate distribution of emergency medical services ([21], [25], [46], [59], [77], [83], [85]).

B. The Structure of Moral Imperatives in Resuscitation

As a rule, there is no moral obligation to attempt the impossible. This generalization, referring equally to attempts to restore life after death and to other efforts in medicine, has been invoked primarily to justify the institution of Do Not Resuscitate (DNR) orders: "Physicians have no obligation to provide futile or useless treatment. A DNR order is appropriate when further treatment is futile, so that successful CPR would only prolong the process of dying" [50]. This sentiment has been echoed with slightly different emphasis in a number of different contexts.

Only one part of the story is represented by these statements, however. Resuscitation has been attempted even when it seemed futile perhaps because, strictly speaking, there *is* an important and deep-seated ideal to further science and to cheat death by quixotically attempting the impossible ([13], [50]). On its constructive side, this vision has been quite inspiring ([58], [65]). On its destructive side, it has been blamed for haphazard, unwanted, and, at times, cruel attempts at futile resuscitation ([27], [28], [32], [61], [65]). Nevertheless, resuscitation is only a *prima facie* obligation ([21], [25], [46], [50], [77], [83], [85]).

Prima facie obligations are by definition, however, not absolute. In specific situations they are weakened or even suspended, and their pre-eminence changes with respect to other obligations. Thus, when an effort is blatantly futile, or when specific treatment for a uniformly fatal condition is unknown, the obligation to resuscitate may be weaker than the obligation to comfort and to prevent or alleviate pain. As knowledge advances, the relative weight of specific competing obligations may change. The extent of the obligation will depend on what one can reasonably accomplish, with the understanding that "reasonableness" is a highly charged and evolving concept.

The purpose of distinguishing between 'revival,' 'resuscitation,' and 'resurrection' is to emphasize the different obligations inherent in the definition of each concept. The obligation to revive is virtually absolute, yet there is a only a *prima facie* obligation to resuscitate. Is there an obligation to resurrect? Not in the way the concept is understood here. To the extent that resurrection is "miraculous", it cannot be expected to succeed, and it is difficult to imagine how a duty to perform miracles can be imposed. In the popular mind there is no clear differentiation between resurrection and resuscitation. This confusion may be responsible for some important misapprehensions in the lay press regarding resuscitation, death, and the significance of DNR orders.

III. THE DNR DECISION: LEGITIMATE MODIFICATIONS OF THE OBLIGATION

There are two circumstances under which we do not, in general, resuscitate: when we cannot, because the effort is futile or impossible, and when we *choose* not to, on the basis of specific criteria. While the *ability* to resuscitate can be studied empirically, the *choice* of whether or not to resuscitate must be considered on ethical grounds. The first deals with what we *can* do: the second, with what we *should* do. We have become accustomed to wide latitude in *choosing* not to resuscitate. We may choose, in response to a patient's wishes, not to resuscitate when death is awaited, or at least is not unwelcome. We ask only to be certain that our perception of these wishes is accurate, and that the patient is intellectually competent in formulating his directives: a strong argument can be made for a universal imperative to prepare advanced directives in case of severe illness or accident, in order to facilitate this process. With these provisos in mind, the patient's desires *should* obtain, and other concerns – the family's opinion, for example – will assume a decidedly secondary role. What we should do when a patient's desires are neither expressed nor apparent is much more difficult to establish. It is generally accepted that such patients **are** to be resuscitated for compelling ethical and legal reasons ([13], [21], [50], [65]). For completeness, it is necessary to raise the problem of the hopelessly ill and mentally incompetent patients who issue impossible directives (while in full command of their intellectual faculties) to be resuscitated and maintained.

Is there, for example, a similar obligation to resuscitate a 90-year-old man riddled with cancer, a 72-year-old woman after her first myocardial infarction, a 68-year-old man after his 5th myocardial infarction, who now shows signs of electro-mechanical dissociation (a fatal arrhythmia if untreated, generally signifying severe damage to the heart), and a 5-year-old child who fell through the ice while skating? Obviously not. The major considerations that bear on this question are worthy of further discussion.

A. FUTILITY

We have already touched on the importance of the concept of futility in modifying the generic obligation to resuscitate. A significant body of data suggests that a subset of patients can be positively identified for whom resuscitation will be ineffectual : Bedell, for example, has shown that patients who do not respond to resuscitation within 30 minutes after cardiac arrest are extraordinarily unlikely to do so. The legitimization of brain death as an acceptable criterion for the [legal] declaration of death is a good example of the practical importance of such data ([9], [11], [17], [20], [51], [52], [62], [81], [86]). The influence of these studies in deciding social policy is growing ([13], [14], [17], [20], [24], [42], [50], [54], [56], [65], [86], [87]). An important part of the real moral debate, however, begins only after these data are collected: how should medical care be influenced by the identification of a population in whom successful resuscitation is a substantial impossibility? How should these studies affect DNR decisions?

The likelihood of affecting resuscitation should not influence the imperative to revive. Revival is "first aid" for a condition or an event that is both self-limited and intrinsically reversible. As we have defined the term, the imperative to revive is virtually absolute (examples to the contrary can be imagined, but they are sufficiently rare to be disregarded for the purposes of this discussion).

For different reasons, the likelihood of effecting resuscitation cannot affect the obligation to resurrect. There is no obligation to attempt to resurrect because resurrection, under ordinary circumstances, is impossible. An important part of moral debate, therefore, relates solely to those restricted situations in which true resuscitation (in the strictest sense of the word) is applicable. These are situations where death threatens, but has not yet occurred; where there is some realistic hope of reversing or controlling the precipitating factor that led to the life-threatening situation; and where any underlying and immediately life-threatening medical problems can also be treated or controlled in some way. When these conditions are not satisfied, there can be no obligation to resuscitate because resuscitation will not be effective: it is not resuscitation that is wanted, but rather some other procedure that might reconstruct the patient's body, and restore his health. Thus, if an irreparable aneurysm of the heart has burst so that the immediate implantation of an artificial or transplanted heart becomes the only hope of survival, the implantation of the substitute organ cannot be subsumed under the aegis of 'resuscitation': any obligation to carry out such a procedure derives from some other place. This type of situation does not enter the province of DNR decision making, though admittedly, many philosophical considerations are common to both. Part of the difficulty in generating DNR policy with respect to the elderly devolves from the difficulty of deciding whether what a given patient requires is resuscitation, revival, or some more fundamental (and generally unattainable) biologic repair.

Death in patients with incurable illnesses is not subject to the same burden of proof as unexpected or accidental death. One is willing to accept the appearance of death at face value because, in most such circumstances, death is neither unbidden nor unfair. Not every incurable illness is the same, however, and even the most incurable condition can be favored by periods of relative well-being. Incurability *alone* does not suffice to modify the imperative to resuscitate, therefore, but when the event immediately precipitating a patient's physiologic decompensation is an irreversible complication of the underlying condition, the futility of the effort (not the incurability of the disease) again modifies any obligation to resuscitate.

The *healthy* 90-year-old patient with his first myocardial infarction is no less worthy of resuscitation than the 45-year-old with his first, but subsequent complications in *either* case may modify this obligation or suspend it altogether. Thirty minutes after CPR has been initiated, one would be under a stronger obligation to continue efforts on the 45-year-old than on the 90-year-old, and, in equivalent circumstances, one would persist even longer in attempting to resuscitate the 5-year-old child who fell through the ice. These decisions are only made on the basis of information regarding *the likelihood of success*, and not on the basis of age or some utilitarian formula.

At the Beth Israel Hospital in Boston, the obligation to resuscitate was modified in the face of "imminent death" from "irreversible disease" [65]. This argument for "reasonable" DNR policies was based on the implicit understanding that the futility of resuscitation in some situations could be predicted, that patients and families could be included in the decision making process, and that "routine" medical care (other than resuscitation) would not be compromised by deciding empirically that resuscitation was, in fact, futile and *therefore* unnecessary.

B. Age

Biologic – as distinguished from chronologic – age serves as a prognostic marker for therapeutic futility in specific circumstances. Above a certain chronologic age, the likelihood of substantial recovery from certain neurological events diminishes rapidly ([51], [52], [81]), and cannot be improved by treatment. Underlying multiple system failure is often latent and quickly precipitated by even minor illness. As a rule, virtually any illness will be more severe in the elderly ([9], [20], [28], [42], [86]). Insofar as any patient with multiple system disease is hypersusceptible to minor setbacks, Epstein was led to conclude that "Failure of the resuscitation attempt was predicted not by age but by disease – circulatory failure, acute stroke, metastatic cancer, uraemia, pneumonia, or sepsis" [23]. Chronologic age, however, has never been successfully correlated with survival after resuscitation, regardless of whether "survival" is defined in hours ([9], [27]), weeks [86], or discharge home. This observation has been substantiated by virtually every major study or policy paper on resuscitation or medical care of the aged in the English language ([19], [20], [21], [23], [27], [35], [42], [50], [65], [86], [87]). This may be part of the reason why the Neasden Memorandum was so shocking.

Although *biologic* age can serve as one of *several* indices which define a population at equal risk of catastrophic illness and futile intervention, the use of *chronologic* age in this way is far more difficult to justify.

C. Dementia

Severe dementia has been proposed as a justification for limiting both resuscitation efforts specifically and the intensity of medical ministrations as a whole. Some important admonitions must be noted. Dementia is a condition with many causes rather than a specific diagnosis. Although true dementia is generally regarded as incurable, illness, bereavement, or stress can unleash a major depression that can mimic rapidly progressive dementia ([30], [33], [45], [88]) (drug reactions can produce similar effects). Such reactive or involutional depressions often respond to treatment, with dramatic reversals of emotional and intellectual debility. As regards true dementia, the treatment of almost any underlying systemic condition can often improve mental function. The literature reflects a cycle in which senescence predisposes to a presumption of senility. Both these labels, "age" and "senility," subject a patient to the substantial risk of benign neglect "out of kindness," to spare the patient the putative suffering of "batteries of investigations" that "cannot be cost beneficial" [20]. This attitude incorrectly presupposes the futility of any effort to improve mental function, and confuses two concerns alluded to earlier: the economic calculus of cost-effectiveness and the paternalistic calculus of projected suffering. This attitude takes its origin from two errors: first, the burden of the patient's age and debility are so overwhelming that the importance of restoring function is lost in the impossibility of restoring *total* health; second, the maxim "first, do no harm" (actually "if you cannot cure, first, do no harm") is misinterpreted to mean "do nothing if you cannot cure."

It has been amply demonstrated that chronologic age does not define mental competence. On the other hand, one is not infrequently left with a situation in which a patient is "slightly demented": mental dysfunction has not yet reached the point of global dementia, but one can never be quite sure whether the patient's comprehension is complete, and whether his or her expressed desires accurately reflect the inner will. The first priority is to attempt to improve mental functioning. Failing this, or failing the receipt of a satisfactory instruction from the patient or his proxy, some form of advocacy must be assured [15].

There is little data available on how frequently "pure dementia" is found in the aged in the absence of any significant somatic disease. True Alzheimer's disease – presenile dementia – may be one disease in which this occurs. Pure dementia must be considered the equivalent of severe single system failure. In the absence of additional instructions we cannot presume to make fair or reliable quality of life decisions. The situation in which the aged patient is despondent and begs to die, and the situation in which the aged patient appears to have no comprehension of the immediacy of his fate, are equally burdensome. If revival is necessary it should be provided, much as it would be in a child with Down's syndrome. Resuscitation to "test" death, however, need not be carried out, and when death threatens it need not be impeded, given the safeguards of appropriate advocacy. A decision tree based on similar policies has been tested and found useful by one institution that "stages" the emergency treatment of potentially life-threatening events in the aged [21].

It is unrealistic to expect dementia to be a neutral condition having little effect on the professional medical staff. The problem arises in hidden assumptions woven into putatively transparent discussions of the medical care of the aged, e.g., "It is the responsibility of the physician to prolong and improve the quality of life. It is our job to treat patients in a holistic, humanistic manner. *But how do our role and function change when the patient is severely demented?*" ([70], emphasis mine). The unfortunate combination of dementia, disease, and aging puts the elderly patient at risk of insufficient, rather than of over-exuberant medical attention. Whether dementia and multiple system disease are justifiable components of the DNR decision will require further analysis before these factors can be incorporated into health care policy.

D. Advanced Directives and Personal Rights

Because of the fundamental importance of determining the patient's wishes in making DNR decisions, an extensive literature on how to elicit expression of these wishes has evolved. The stated concern is typically to allow the patient the option of *declining* resuscitation, thereby emphasizing the right to decline care and to control one's time, place, and manner of death ([2], [6], [10], [13], [14], [15], [20], [21], [27], [30], [32], [33], [42], [50], [56], [65], [86], [87]). It is tacitly assumed that when there is disagreement between patient and medical staff, it will take the form of the patient's relinquishing, and the staff's encouraging, any further call on life. The reverse, wherein the staff gives up hope or interest while the patient hangs on, is just as important – particularly in geriatrics – but rarely discussed. Very often, this latter situation finds expression in action ("slow code") rather than words. One wonders whether physicians speak differently to patients from whom they expect accord than to others. The right to decline resuscitation is more strongly defended than the right to demand it.

Given this emphasis on avoiding "wrongful," or "unwanted living," the courts have been fairly liberal in protecting the medical profession from liability in "wrongful death" actions brought by plaintiffs following DNR decisions. "Wrongful *life*" actions as a consequence of *unwanted* resuscitation, have not, on the whole, been prominent ([22], [32]). The

courts have also taken on aspects of the physician's traditional benefi-
cent (and paternalistic) prerogative of making decisions on behalf of less
than fully competent patients. This involvement has not been uniformly
welcomed, despite the general support for DNR procedures in most
jurisdictions. Perhaps because many details regarding the interpretation
and execution of these orders remain unexplored, physicians have
assumed a decidedly defensive posture, more for their own protection
than for ethical considerations. Thus, it has been recommended that
"this decision be openly discussed by physicians and families; that
witnesses and written documentation of the process and the decision be
recorded; and that where significant disagreement or questions arise,
the courts be brought into the decision making process" [56].

In the belief that many people hold strong preferences concerning
how they would wish to be treated when catastrophically ill, legislation
legitimizing the provisions of advanced directives such as "living wills"
has been introduced (although "living wills" are a specific form of
advanced directive, the term living will is used generically). Initially,
living wills were intended to facilitate organ donation. More recently,
they have become attractive to those who would consider limiting *in
advance* the resuscitative efforts to which they would be subjected when
incapable of making intelligent choices. Living wills apply to two different categories of illness: 1) known chronic or degenerative conditions
where the need for continuing choices concerning the type, intensity, or
direction (*e.g.*, cure *vs.* palliation) of treatment can be anticipated; 2)
unexpected catastrophic conditions that destroy the possibility of making decisions or expressing one's will. Do these directives have the
ability to modify the obligation to resuscitate? In every jurisdiction, at
the time of this writing a patient must be certified as terminally ill and
incapable of expressing his wishes before the provisions of a living will
can be enacted: the state's compelling interest in having the physician
treat the patient only collapses when treatment is futile. Insofar as DNR
decisions are concerned, therefore, advanced directives are moot because they can take effect only after the obligation to resuscitate has
already weakened. Some states have passed legislation enabling a
"durable power of attorney" (DPA), permanently transferring the
power to make decisions on the part of one individual to another,
should that individual lose legal competence. The DPA legislation is
capable of serving both categories of illness outlined above, and may
eventually have a greater influence on DNR policy than the living will.

One question that has not been addressed is how to respond to
requests not to limit the intensity of medical care, even when that care is
futile. This request is often heard on hospital wards, and becomes the
source of much anguish, frustration, and anger.

IV. SUMMARY AND CONCLUSIONS

In the clinical setting, physicians offer four reasons for limiting treatment: futility, patient choice, quality of life, and economic costs [51].
The first two are more or less objective: the second are highly subjective
and require a difficult cost-benefit evaluation. The first two reasons,
safe, relatively unambiguous, and fairly consistent with the Beth Israel
formulation, are more frequently cited; but the second two, in many
ways more critical to the coherence of health care delivery systems, are
tacitly more influential because they have the appearance of objectivity.
They appeal to a "greater good" on a macro-economic scale. In actual-
ity, these reasons are invoked mostly when the first two fail to provide a
satisfactory excuse for allowing a patient to die when the medical staff
feels he should. Can these criteria for DNR decisions find any additional
justification?

It is very difficult to collect meaningful data on quality of life and
cost-benefit issues ([4], [12], [29], [54], [60], [67]), particularly with
respect to the resuscitation of the aged ([1], [9], [27], [38], [55], [82]).
On the one hand, the number of patients surviving after resuscitation is
remarkably low, so that the realistic *economic* strain imposed by "uncritical resuscitation" is potentially quite limited. More often than not,
resuscitation illuminates the fact that apparent death is virtually death in
fact. On the other hand, the cost of care during end-stage illness can
easily exceed the cost at any other time ([9], [17], [21], [27], [35], [51],
[52], [81]). Issues of cost in the distribution of medical care are seen in
contradictory ways. American society takes pride in providing the
"best" medical care *regardless* of cost, despite almost constant mention
in the media and in the legislatures of the need to restrict the cost of
medical care. The systematic identification and ordering of priorities in
medicine is sometimes taken to contradict this point; the planned
distribution of medical care has come to be viewed simultaneously as
morally necessary and ideologically (if not morally) difficult. Although it
is beyond the scope of this essay to explore the dimensions of either
viewpoint, it is obvious that such questions have far-ranging implications for the elderly.

The monetary and social costs of honoring the obligation to prolong and to save life must always be balanced against the moral and philosophical costs of suspending it. The importance of economic and utilitarian factors cannot be denied. But because decisions based on incomplete data with respect to these factors can cause irretrievable damage both to individual patients *and* to society, their role should probably be abridged at the present time. When cost-benefit analyses are proposed, their ramifications must be explored to their fullest extent. It is one thing to propose a cost-benefit analysis in deciding where to build a hospital and how many intensive care beds to install: it is another to decide that *this* life at *this* time should not be prolonged for economic reasons.

Any patient has the right to decline medical care within the confines of his competence. A competent patient has the right to decline resuscitation and, arguably, the right to decline revival as well. Given the frequently "non-medical" nature of revival, one could argue that revival should be carried out even against the patient's expressed desires.

The most important consideration, then, in formulating DNR policies, both in general and for the aged, is individualization of care in the decisionmaking process. To the extent that generalizations are possible the following principles are suggested:

1. Revival should always be attempted.
2. There is no requirement to attempt resurrection.
3. The applicability of "true" resuscitation, though limited, is totally individualized.
4. Whereas chronologic age is not to be determinative in DNR policy formulations, biologic age may be, but only insofar as it modifies the feasibility of resuscitation and, by extrapolation, the obligation to carry it out.
5. The wishes of the patient are pre-eminent, and every effort must be made to cultivate and elicit their expression.
6. When no definite expression of a patient's wishes is available, or when unforeseen variables intervene, the obligation to restore and to prolong life must predominate.

In sum, in *most* situations where one should not attempt to resuscitate, it should be because one cannot, rather than because one chooses not to. In *most* situations where we are uninstructed by the patient, or in any way uncertain of the patient's specific wishes, we are obliged to attempt to restore life. As more data are obtained, a better definition will evolve of the population in whom resuscitation will fail, and as technical advances overcome such failings, the possibility of successful attempts to resuscitate and resurrect will necessarily change.

Walter Reed Army Medical Center
and the Kennedy Institute of Ethics
Washington, D.C.

DISCLAIMER

The opinions expressed in this essay represent the personal views of the author, and should not be construed as representing the policies of the Department of Defense or the Department of the Army.

ACKNOWLEDGEMENT

I wish to acknowledge the assistance of Jae and Jonathan Roosevelt, and Linda Rabinowitz Dagi, whose critical readings of successive versions of this essay contributed both form and substance to its development.

BIBLIOGRAPHY

1. Abramson, N. S., Meisel, A. and Safar, P.: 1981, 'Informed Consent in Resuscitation Research', *Journal of the American Medical Association* **246**, 2828–2830.
2. Angell, M.: 1984, 'Respecting the Autonomy of Competent Patients', *New England Journal of Medicine* **310**, 1115–1116.
3. Anonymous: 1732, *Notes From the Dead: In the Revival of Anne Green, Hanged at Oxford in 1650. With Near 40 Ingenious Poems on the Subject in Latin English, etc. by the Prime Wits of that University*, Phoenix Brittanicus, Vol. I; cited in [85] (W. G. A. Robertson).
4. Avorn, J.: 1984, 'Benefit and Cost Analysis in Geriatric Care. Turning Age Discrimination into Health Policy', *New England Journal of Medicine* **310**, 1294–1301.
5. *Babylonian Talmud*, Tractate Shabbat, 152b.
6. Baer, L. S.: 1978, 'Nontreatment of Some Severe Strokes', *Annals of Neurology* **4**(4), 381–382.
7. Beecher, H. K.: 1968, 'Ethical Problems Created by the Hopelessly Unconscious Patient', *New England Journal of Medicine* **278**, 1425–1430.
8. Beecher, H. K.: 1969, 'Procedures for the Appropriate Management of Patients Who May Have Supportive Measures Withdrawn', *Journal of the American Medical Association* **209**, 405.

9. Bedell, S. E., Delbanco, T. L., Cook, S. F. *et al.*: 1983, 'Survival After Cardiopulmonary Resuscitation in the Hospital', *New England Journal of Medicine* **309**, 569–575.

10. Bedell, S. E. and Delbanco, T. L.: 1984, 'Choices About Cardiopulmonary Resuscitation in the Hospital. When Do Physicians Talk with Patients?' *New England Journal of Medicine* **310**, 1089–1093.

11. Black, P. McL.: 1980, 'From Heart to Brain: The New Definitions of Death', *American Heart Journal* **99**(3), 279–81.

12. Campion, E. W.; Mulley, A. G.; Goldstein, R. L., *et al.*: 1981, 'Medical Intensive Care for the Elderly. A Study of Current Use, Costs, and Outcomes', *Journal of the American Medical Association* **246**, 2052–2056.

13. Cassem, N.: 1980, 'When Illness is Judged Irreversible: Imperative and Elective Treatments', *Man and Medicine* **5**(3), 154–166.

14. Cassem, N. H.: 1980, 'Consultation to Continue or Stop Treatment Measures in Irreversible Illness', *Advances in Psychosomatic Medicine* **10**, 119–131.

15. Dagi, T. F.: 1974, 'The Ethical Tribunal in Medicine', *Boston University Law Review* **54**(2), 269–277.

16. Dagi, T. F.: 1978, 'Cause and Culpability', *Journal of Medicine and Philosophy* **1**(4), 349–377.

17. Davies, M. J. and Thomas, A.: 1984, 'Thrombosis and Acute Coronary Artery Lesions in Sudden Cardiac Ischemic Death', *New England Journal of Medicine* **310**, 1137–1140.

18. DeBard, M. L.: 1980, 'The History of Cardiopulmonary Resuscitation', *Annals of Emergency Medicine* **9**, 273–275.

19. Editorial: 1982, 'Cardiac Resuscitation in Hospital: More Restraint Needed?' *The Lancet* **1**, 27–28.

20. Editorial: 1982, 'Death in Old Age', *The Lancet* **2**, 477.

21. Eisenberg, M. S., Hallstrom, A. and Bergner, L.: 1982, 'Long Term Survival after Out of Hospital Cardiac Arrest', *The New England Journal of Medicine* **306**, 1340–1343.

22. Eisendrath, S. J. and Jonsen, A. R.: 1983, 'The Living Will. Help or Hindrance? *Journal of the American Medical Association* **249**, 2054–2058.

23. Epstein, F. H., Bedell, S. E., and Delbanco, T. L.: 1983, 'Cardiopulmonary Resuscitation of Old People', [letter] *The Lancet* **2**, 794.

24. Evans, K. G.: 1981, '"No Resuscitation" Orders – An Emerging Consensus', *Canadian Medical Association Journal* **125**, 892–896.

25. Federal Emergency Medical Systems Act: 1980, 42 *USCA* sec 300a–300d–9 (1980 Suppl).

26. Fletcher, G. P.: 1967, 'Prolonging Life', *Washington Law Review* **42**, 999–1016.

27. Fuesgen, I. and Summa, J.-D.: 1978, 'How Much Sense is There in an Attempt to Resuscitate an Aged Person?' *Gerontology* **24**, 37–45.

28. Gillick, M.: 1980, 'The Ethics of Cardiopulmonary Resuscitation: Another Look', *Ethics in Science and Medicine* **7**, 161–169.

29. Gillick, M.: 1984, 'Is the Care of the Chronically Ill a Medical Prerogative?' *New England Journal of Medicine* **310**, 190–191.

30. Graham, J. and Livesley, B.: 1983, 'Dying as a Diagnosis: Difficulties of Communication and Management in Elderly Patients', *The Lancet* **2**, 670–672.

31. Hartikainen, M., Cozanitis, D. A. and Heikkila, J.: 1982. 'Resurrection', *Southern Medical Journal* **75**(10), 1301.
32. Hashinmoto, D. M.: 1983, 'Structural Analysis of the Physician Patient Relationship in No-Code Decision Making', *Yale Law Journal* **93**, 362–383.
33. Haug, M.: 1978, 'Aging and the Right to Terminate Medical Treatment', *Journal of Gerontology* **35**(4), 586–591.
34. Henry, J. B. and Smith, F. A.: 1980, 'Estimation of the Postmortem Interval by Chemical Means', *American Journal of Forensic Medical Pathology* **1**, 341–347.
35. Hershey, C. O. and Fisher, L.: 1982, 'Why Outcome of Cardiopulmonary Resuscitation in General Wards is So Poor', *The Lancet* **1**, 31–34.
36. Hunter, J.: 1776, 'Proposals for the Recovery of People Apparently Drowned', Phil Reans 66, 412; cited in *Deciding to Forego Life-Sustaining Treatment*, [63], p. 232.
37. Jakobovits, I.: 1959, *Jewish Medical Ethics*, Bloch Publishing Company, New York.
38. Kapp. M. B.: 1983, 'Age and Mental Incompetence', [letter] *Annals of Internal Medicine* **98** (Part I), 669–670.
39. Key References in Cardiopulmonary Resuscitation: 1982, Part I, *Circulation* **66**(4), 898–900; continued in Part II, *Circulation* **66**(5) 1133–1135.
40. II *Kings*: 4:31–37.
41. Knight, B.: 1983, 'A Comparative Survey of the Medico-Legal Aspects of Death in Europe', *Medicine and Law* **2**(2) 137–156.
42. Kohn, R. R.: 1982, 'Cause of Death in Very Old People', *Journal of the American Medical Association* **247**, 2793–2797.
43. Kouwenhoven, W. B., Jude, J. R. and Knickerbocker, G. G.: 1960, 'Closed-chest Cardiac Massage', *Journal of the American Medical Association* **173**, 1064–1067.
44. Lamb, D. and Easton, S. M.: 1982, 'Philosophy of Medicine in the United Kingdom', *Metamedicine*, **3**, 3–34.
45. Levenson, S. A., List, N. D. and Zaw-Win, B.: 1981, 'Ethical Considerations in Critical and Terminal Illness in the Elderly', *Journal of the American Geriatrics Society* **29**(12), 563–567. See Table II in particular.
46. Lewis, B.: 1983, 'Death in the First Ten Minutes', *British Medical Journal* **286**, 1768–1769.
47. Lidz, C. W., Meisel, A., Osterweis, M., *et al.*: 1983, 'Barriers to Informed Consent', *Annals of Internal Medicine* **99**, 539–543.
48. Linko, K., Honkavaara, P. and Salmenpera, M.: 1982, 'Recovery after Discontinued Cardiopulmonary Resuscitation', *The Lancet* **1**, 106–107.
49. Lo, B. and Jonsen, A. R.: 1980, 'Clinical Decisions to Limit Treatment', *Annals of Internal Medicine* **93**, 764–768.
50. Lo, B. and Steinbrook, R. L.: 1983, 'Deciding Whether to Resuscitate', *Annals of Internal Medicine* **143**, 1561–1563.
51. Longstreth, W. T., Diehr, P. and Inui, T. S.: 1983, 'Prediction of Awakening after Out-of-Hospital Cardiac Arrest', *New England Journal of Medicine* **308**, 1378–1382.
52. Longstreth, W. T., Inui, T. S., Cobb, L. A., *et al.*: 1983, 'Neurologic Recovery after Out-of-Hospital Cardiac Arrest', *Annals of Internal Medicine* **98** (Part I), 588–592.
53. Lund, I. and Skulberg, A.: 1976, 'Cardiopulmonary Resuscitation by Lay People', *The Lancet* **2**, 702–704.
54. MacDonell, J. A.: 1981, '"No Resuscitation" Orders', *Canadian Medical Association Journal* **125**, 809–810.

55. Micetich, K. C. and Thomasma, D. C.: 1984, 'The Ethics of Patient Requests in Experimental Medicine', *CA: A Journal for Clinicians* **34**(2), 118–120.
56. Miles, S. H., Cranford, R. and Schultz, A. L.: 1982, 'The Do-Not-Resuscitate Order in a Teaching Hospital. Considerations and a Suggested Policy', *Annals of Internal Medicine* **96**, 660–664.
57. Morris, N: 1958, 'The History of Cardiac Resuscitation', in H. E. Stevenson, Jr. (ed.), *Cardiac Arrest*, C. V. Mosby, St. Louis, pp. 15–31.
58. Morse, M.: 1983, 'A Near-Death Experience in a 7-Year-Old Child', *American Journal of the Diseases of Children* **137**, 959–961.
59. Motro, H.: 1983, 'Medicolegal Aspects of Cardiopulmonary Resuscitation', *Medicine and Law* **2**(2), 103–112.
60. Najman, J. N. and Levene, S.: 1981, 'Evaluating the Impact of Medical Care and Technologies on the Quality of Life: A Review and Critique', *Social Science and Medicine* **15**(f), 107–115.
61. Negovsky, V.: 1982, 'Reanimatology Today: Some Scientific and Philosophic Considerations', *Critical Care Medicine* **10**(2), 130–133.
62. Powner, D. J. and Grenvik, A.: 1979, 'Triage in Patient Care: From Expected Recovery to Brain Death', *Heart and Lung* **8**(6), 1103–1108.
63. President's Commission for the Study of Ethical Problems in Medicine and Biomedical and Behavioral Research: 1983, *Deciding to Forego Life-Sustaining Treatment*, U.S. Government Printing Office, Washington, D.C.
64. Preuss, J.: 1978, *Biblical and Talmudic Medicine*, translated and edited F. Rosner, Sanhedrin Press, New York.
65. Rabkin, M. T., Gillerman, G. and Rice, N. R.: 1976, 'Orders Not to Resuscitate', *New England Journal of Medicine* **295**, 364–366.
66. Ramsey, P.: 1976, 'Prolonged Dying: Not Medically Indicated', *Hastings Center Report* **6**, 14–17.
67. Robertson, G. S.: 1983, 'Ethical Dilemmas of Brain Failure in the Elderly', *British Medical Journal* **287**, 1775–1777.
68. Robertson, W. G. A.: 1935, 'Recovery after Judicial Hanging', *British Medical Journal* (Jan 19), 121–122.
69. Rosenberg, A. J.: 1980, *Mikraot G'dolot: The Book of Kings, Translation of Text, Rashi and Commentary*, Judaica Press, New York.
70. Ross, W. D.: 1930, *The Right and The Good*, Oxford University Press, London.
71. Safar, P.: 1982, 'Reanimatology', *Critical Care Medicine* **10**, 134–136.
72. Schneider, E. L. and Brody, J. A.: 1983, 'Aging, Natural Death, and the Compression of Morbidity – Another View', *New England Journal of Medicine* **309**, 854–856.
73. Shakespeare, W.: 1917, *Romeo and Juliet*, Yale University Press, New Haven, Conn.
74. Sladen, A.: 1984, 'Closed-chest Massage, Kouwenhoven, Jude, Knickerbocker', *Journal of the American Medical Association* **251**, 3137–3140.
75. Soleveichik, A.: 1978, 'Jewish Law and the Time of Death', [letter] *Journal of the American Medical Association* **240**, 109.
76. Sowden, G. R., Robins, D. W. and Baskett, J. F.: 1984, 'Factors Associated with Survival and Eventual Cerebral Status Following Cardiac Arrest', *Anesthesia* **39**, 39–43.
77. St. Louis, P., Carter, W. B. and Eisenberg, M. S.: 1982, 'Prescribing CPR: A Survey of Physicians', *American Journal of Public Health* **72**(10), 1158–1160.

78. Anon.: 1980, 'Standards and Guidelines for Cardiopulmonary Resuscitation (CPR) and Emergency Cardiac Care (ECC)', *Journal of the American Medical Association* **244**, 453–509.
79. Stevenson, I: 1977, 'Research into Evidence of Man's Survival after Death: A Historical and Critical Survey with a Summary of Recent Developments', *Journal of Nervous and Mental Disease* **165**, 152–170.
80. Tendler, M. B.: 1978, 'Jewish Law and the Time of Death', [letter] *Journal of the American Medical Association* **240**, 109.
81. Teres, D., Brown, R. B. and Leneshow, S.: 1982, 'Predicting Mortality of Intensive Care Unit Patients: The Importance of Coma', *Critical Care Medicine* **10**(2), 86–95.
82. Thomasma, D. C.: 1983, 'Beyond Medical Paternalism and Patient Autonomy: A Model of Physician Conscience for the Physician–Patient Relationship', *Annals of Internal Medicine* **98**, 243–248.
83. Thomson, R., Hallstrom, A. and Cobb, L.: 1979, 'Bystander-Initiated Cardiopulmonary Resuscitation in the Management of Ventricular Fibrillation', *Annals of Internal Medicine* **90**, 737–740.
84. Veith, F. J., Fein, J. M., Tendler, M. D., *et al.*: 1977, 'Brain Death I. A Status Report of Medical and Ethical Consideration', *Journal of the American Medical Association* **238**, 1651–1655; and [in continuation] 'Brain Death II. A Status Report of Legal Considerations', *Journal of the American Medical Association* **238**, 1744–1748.
85. *Vt. Stat. Anno.*, ch. 12, sec 519 (1981).
86. Wagner, A.: 1984, 'Cardiopulmonary Resuscitation in the Aged', *New England Journal of Medicine* **310**, 1129–1130.
87. Wanzer, S. H., Adelstein, S. J., Cranford, R. E., *et al.*: 1984, 'The Physician's Responsibility Towards Hopelessly Ill Patients', *New England Journal of Medicine* **310**, 955–959.
88. Wilson, P.: 1982, 'Anxiety and Depression in Elderly and Dying Patients', *Medical Clinics of North America* **66**(5), 1011–1016.

SECTION III

SELF-DETERMINATION IN LATE-LIFE DEPENDENCY

MOLLY REES GAVIN AND GAYLE KATAJA

SELF-DETERMINATION IN LATER LIFE: CASE STUDIES IN GERIATRIC CARE

Yes I am old – my strength declines
And wrinkles tell the touch of time.
Yet I might fancy these the signs
Not of decay, but Manhood's prime
For all within is young and glowing
Spite of old age's outward showing.

Yes, I am old, experience now
That best of guides hath made me sage,
And thus instructed I avow
My firm conviction that old Age
Of all our various terms of living
Demands our warmest, best Thanksgiving.

Isaac Bell, 1857.

The trans-disciplinary approach is a crucial factor in determining appropriate care for older adults. Connecticut Community Care, Inc. is a statewide private, non-profit agency; the only licensed case management agency in Connecticut, with eight years of experience in using an interdisciplinary team. The function is to provide assessments of older adults, coordination of all services, benefits and entitlements, and monitoring of services correlated to the clients' response. Connecticut Community Care contracts with existing agencies to provide a variety of services such as adult day care, home health care, counseling, companion, and meals. The agency facilitates access and entry into a complex health and social service system, and assists frail elderly clients in obtaining maximum benefits from the system. The targeted population is defined as "high risk, frail elderly" because of multiple problems which warrant the response of an interdisciplinary team. Referrals are taken from any source and assessments of social, financial and health care needs are conducted with the client regardless of the client's location in the health care continuum.

The following cases chronicle the lives of two older adults who were

129

Stuart F. Spicker, Stanley R. Ingman, and Ian R. Lawson (eds.),
Ethical Dimensions of Geriatric Care, 129–136.
© 1987 *by D. Reidel Publishing Company.*

served by Connecticut Community Care, Inc. The two people demonstrate some of the dilemmas (practical and ethical) confronting providers
of care to the elderly.

Mr. John Ansen was referred to Connecticut Community Care, Inc. by a staff person of
the City Division of Aging. Mr. Ansen was a patient at a local skilled nursing facility and the
referring person from the Division of Aging had heard rumors in the community that Mr.
Ansen wanted to go home. A telephone call was placed to the skilled nursing facility and
the social worker from the facility stated that yes, Mr. Ansen wished to go home, but the
medical doctor did not want him discharged. Arrangements were made and an assessment
was conducted shortly thereafter.

Mr. Ansen was an 85-year-old widowed male originally from Estonia. He was admitted
to a local acute care hospital last November for treatment of increasing weakness and
edema of his lower extremities. Mr. Ansen had a history prior to the hospitalization of
recurring dizziness and difficulty with ambulation. A diagnosis of cerebral arteriosclerosis
was established and Mr. Ansen had been maintained on aspirin daily.

He remained hospitalized approximately one month and diagnoses listed on his discharge summary included anemia, splenomegaly secondary to cirrhosis, esophageal vari-
ces, arteriosclerotic heart disease, congestive heart failure, cerebral vascular disease,
questionable transient ischemic attack, and bronchitis. Significant past medical history
included a total laryngectomy for carcinoma of the larynx in 1969 with a subsequent
tracheostomy and a cholecystectomy also in 1969.

Up until his hospitalization in November, he had been cared for at home by his wife.
Unfortunately, Mrs. Ansen had died the previous summer and no relatives existed. Mr.
Ansen belonged to an Estonian church in the area and had a large support system of
church members. One such member of the congregation, Mrs. Stone, worked as a nurses'
aide in a local skilled nursing facility and it was to this facility he was transferred after his
stay in the acute care hospital. Mr. Ansen was a college graduate who taught economics at
a local university. He was fluent in six languages, including Russian, Finnish, French, and
German. His most recent employer was a large department store where he worked as an
executive in the finance department.

Mr. Ansen was interviewed while still a patient in the skilled nursing facility. He had a
great deal of trouble speaking because of the tracheostomy and also had difficulty hearing,
so communication was done via note writing. His first words were: "I want to go home."
He went on to explain that he felt a prisoner, he found the care at the facility to be
adequate, but non-personal and the caregivers rude and at times offensive. He was tired of
being regimented and treated like a child. Over and over again he repeated, "I want to go
home," and finally began sobbing as he spoke.

Mr. Ansen was totally alert and oriented. He scored 10 of 10 questions asked on his
mental status questionnaire. Judgment and reasoning were sound. Mr. Ansen was sitting
in a wheelchair throughout the initial interview. According to the medical record, he
ambulated with a walker twice a day. Pitting edema of both legs was evident. An open
tracheostomy stoma existed covered by black mesh material. Apical rate was 78, S1 and
S2 were normal, a grade II systolic ejection murmur was present. Respirations were 18
and regular. Diarrhea had been a problem for several days and he had been incontinent of
stool several times. He also had been non-compliant with the prescribed sodium restricted

diet. He denied worrying or nervousness, but admitted to being always depressed since admission to the skilled nursing facility. He cried repeatedly while discussing his place-ment. He also admitted to feeling that sometimes his life wasn't worth living, but had no plans to end it. When asked to rate his health as good, fair, or poor, he chose fair. Mr. Ansen admitted to being stubborn and independent, but he agreed to cooperate with Connecticut Community Care, Inc.

Mr. Ansen's funds were limited and a Medicaid, Title 19, application had been filed with the state. Because of his condition, it was expected that Medicare, Title 18, would be the first funding source for home health care and later Medicaid with Connecticut Community Care, Inc. funds supplementing. The following care plan was proposed to the client:

1. Mr. Ansen would remain institutionalized until his gastrointestinal problems subsided.
2. Funding sources for the care plan would be secure before Mr. Ansen was discharged.
3. Connecticut Community Care, Inc. would contact his friend and church member, Mrs. Stone, who worked at the skilled nursing facility to see if she could supplement care arranged via the traditional home health care system.
4. Home delivered meals, seven days a week, to be coordinated via two meal sites.
5. Home Health Aide two hours in the morning and two hours in the evening seven days a week to provide personal care.
6. A lifeline unit (personal emergency response system) would be rented for the client to summon assistance if he had difficulty using the telephone.
7. Nursing supervision of his medications and cardiovascular status to be orchestrated via a home health care agency.
8. Physical therapy to evaluate ambulation.
9. Ongoing case management would be provided by Connecticut Community Care, Inc.

Following the initial assessment, a meeting was held with the director of nursing, director of social service and Mr. Ansen's charge nurse. All three disagreed with the proposed plan and felt the doctor would discharge him against medical advice.

They all felt Mr. Ansen needed 24-hour skilled nursing. Yes, they agreed that he was totally competent to make decisions, but they felt any home care would fail. Initial contact with the physician was pessimistic. Reluctantly, he agreed to sign orders for home care services.

Contact with his friend, Mrs. Stone, was more positive. She was in complete agreement with the proposed plan and offered to supplement the care whenever needed. She felt that he had withered in the facility and felt he would surely die if he did not go home. Mr. Ansen was approved for Title 19, Medicaid. He continued to participate in physical therapy and grew stronger although he still spent most of his time in a wheelchair. Because of the reluctance by the physician and the skilled nursing facility regarding home care, a second assessment was done by another Connecticut Community Care, Inc. case manager and the results concurred with findings from the first assessment.

All pieces of the care plan were in place. A Home Health Aide had been found through a local agency who lived in the same apartment complex as Mr. Ansen. She would work two hours in the morning, (8:00 a.m. to 10:00 a.m.) and two hours in the evening, (6:00 p.m. to 8:00 p.m.). Nursing supervision and physical therapy were coordinated through the same agency. Home delivered, 2 gram sodium meals were coordinated through two providers, one responsible for Monday through Friday, one responsible for Saturday and Sunday. A life line unit (personal emergency response system) was not available immediately, but Mr. Ansen was placed on a priority waiting list. His friend, Mrs. Stone, agreed to stop over a few times during his initial days at home. He arrived home as scheduled and wept as he entered the apartment.

Saturday evening, he developed difficulty breathing, the first difficulty he had had with his tracheostomy in 15 years. An ambulance was summoned and he refused to go back to the hospital. His tracheostomy was suctioned and he felt much better. He again that night developed difficulty and he was transported back to the acute care hospital with a diagnosis of plural effusion. A few days later a compression fracture of L_1 was discovered. Two weeks into the hospitalization, Mrs. Stone called to say Mr. Ansen had given up and his spirit was broken. A few days later he improved and requested home care again and the physician and nurses suggested skilled nursing facility care. A compromise was reached and Mr. Ansen was sent to a rehabilitation unit of the hospital. His main problem was inability to transfer. The plural effusion and compression fracture had resolved. Again he became motivated and somewhat successful with therapy. Discharge plans were discussed and coordinated with the hospital and a tentative discharge date was agreed upon. Identical home care plans were set up for Mr. Ansen as in the discharge from the skilled nursing facility.

A 92-year-old Swedish widow telephoned Connecticut Community Care, Inc. after a two-month placement in a nursing facility. Mrs. Harper had been consistently unhappy in the nursing home and upon recommendation by a local geriatric psychiatrist, she was given the name and number of the agency by the social worker in the facility in order to "test" her motivation to return to community living and to gather information re: home care. Mrs. Harper's placement in the nursing facility immediately followed an acute care hospital admission for "dizziness . . . secondary to poor nutrition." According to her daughter, "she was so run down from not eating properly."

Prior to the assessment interview, a Connecticut Community Care, Inc. social worker contacted Mrs. Harper's family to evaluate their ability to assist her. Her daughter and son-in-law, in a series of telephone interviews, indicated that since the death of Mrs. Harper's husband in 1969 . . . "mother has gotten everything she has requested from the family" . . . housework, meals, shopping, and heavy laundry. This particular daughter is employed outside the home. Mrs. Harper's son-in-law is wheelchair bound due to multiple sclerosis. Another daughter is on dialysis. Mrs. Harper's oldest son has "bouts" with cancer. The other son is married to a woman with "heart trouble." Family felt that heretofore they would be unable to provide Mrs. Harper with any of the aforementioned concrete tasks other than laundry.

Multiple attempts had been made in the past to involve Mrs. Harper with home care services – but she was extremely resistant.

The family acknowledged legitimate concern about money. Daughter and son-in-law were aware of the fact that Title 19, Medicaid, would pay for nursing home care should Mrs. Harper exhaust all her financial resources . . . but . . . "who will pay for home care?"

At the time of the Connecticut Community Care, Inc. assessment, Mrs. Harper's medical problems included hyperuricemia, refractory anemia, osteoarthritis and arteriosclerotic heart disease. However, the Director of Nursing and staff nurse reported Mrs. Harper's limited mentation was her most serious problem: severe short-term and some long-term memory impairment. (An example used by nursing staff was that after being in the nursing facility for over one month – Mrs. Harper was still sometimes disoriented re. such things as which bed was hers.) Nursing staff did not feel Mrs. Harper could be taught to use an in-home emergency call system. Discussion with Mrs. Harper's daughter revealed that the daughter was very concerned about her mother's mentation. The daughter indicated that Mrs. Harper is "unaware of the extent that family is helping her."

During the interview with Mrs. Harper, she denied that she had been hospitalized and denied that family had been helping with meals indicating that she is a "wonderful cook" who "loves to cook." She had worked all her life as a housekeeper. When unable to answer certain specific questions, she said she "doesn't keep track of certain things."

Throughout the course of this lengthy assessment interview, Mrs. Harper consistently reiterated her unhappiness in the nursing facility and her desire to return home. There were three options available:

1. Mrs. Harper could remain in the nursing home, but would be very depressed and continue to have adjustment problems.
2. Mrs. Harper could return home with live-in help (recognizing that good, reliable help is difficult to get; costly; and problems reoccur re. terminations, etc.; making the option extremely unrealistic).
3. Mrs. Harper could return to home with daily (but not live-in) help for meal preparation and medication supervision. Risks included the possibility of Mrs. Harper's being home alone and not being able to handle an emergency.

In subsequent discussions with Mrs. Harper's daughter, the daughter stated she would rather have her mother in the nursing home and depressed than have mother come home with the family accepting the risks and guilt of her being home without 24-hour care. Daughter clarified that she had power of attorney and was applying for conservatorship of person. She valued her mother's safety more than Mrs. Harper's emotional satisfaction. While family pondered alternatives, Mrs. Harper continued to check in with the nursing home social worker and the Connecticut Community Care, Inc. staff person re. "progress" of discharge plans.

Mrs. Harper's nursing facility physician would make no concrete recommendation re. home care. He indicated he had only seen her in an institutional setting and, therefore, couldn't make an assessment re. her ability to function at home. He described her as "very confused."

A geriatric psychiatrist who had evaluated Mrs. Harper twice in the nursing home acknowledged her memory impairment, but felt her judgment and mentation were adequate to sustain community living with appropriate home care planning and support. He felt her judgment was adequate to know how to react in an emergency (i.e., to get out of the house in the event of fire). Contrary to institutional nursing personnel, he thought her mentation was adequate to learn how to use an in-home emergency call system.

The Connecticut Community Care, Inc. social worker made a second visit to Mrs. Harper to share the family's concern about how Mrs. Harper would manage during the time no one was with her.

In a stirring soliloquy, Mrs. Harper shared with the Connecticut Community Care, Inc. social worker her philosophy re. aging . . . "when your time comes, your time comes . . ." She discussed that she was not afraid of catastrophes while she was home alone because she was old and ready to die, but she wanted to be in her own home. She talked about her husband's dying in the 80's in his own home; about her mother's dying at the age of 92 in her own home; saying they were old – ready to die because they had lived fine lives and happy because fhey were in their own homes where they belonged. She said it was her decision to make – she didn't care what her daughter said, she was going home.

As John Derven and associates in an article entitled, "Ethical Considerations in Eldercare" query . . . who will decide . . . with what risks . . . and for which outcomes . . . which (medical) interventions will or will not be used or . . . to put it another way, how will the dead testify as to the fit of their dying?

A preliminary study by Marilyn Smallegan regarding decision making for nursing home admission elicited information regarding 19 patients and revealed that a total of 34 persons were involved in admission decisions for these 19 individuals. Nine of the 19 patients described themselves as having *some* responsibility for decision making, but the second highest number of decision makers (seven in both categories) were adult children and social workers. Clearly, in the case of Mrs. Harper, the children's decision would not likely have dovetailed with what their mother wished.

As described in a popular magazine article, "When to Intrude in Your Parents Lives?", whose decision is it to take a chance on living or dying? The parent's . . . or the child's? ·

The role of social workers in this decision-making capacity is not surprising in light of the role played by so many social workers in acute care hospitals. What is somewhat alarming in this regard is a rather disheartening study attempting to "validate" professional judgment decisions regarding patient care planning. Data underscored the low reliability among virtually all professionals participating in the study.

What was even more disheartening was the obvious evidence supporting the fact that the professional's *place of employment* more so than professional training was a strong influence in decision making. Nursing home assessors were simply more likely to see clients as needing nursing home care than were home care assessors. Data indicated greater consistency in ratings by professionals working in the *same setting* than among those of the same professional group. In the case of both Mr. Ansen and Mrs. Harper, institution-based providers of care were not supportive of the older adults' goal for home care. Who decides?

These same 19 patients indicated that 26 people were used for consultation purposes in making this difficult decision in addition to the 34 "decision makers." Ten of these 26 were physicians. Mr. Ansen's physician did not wholeheartedly support the home care plan.

In the case of Mrs. Harper, the physician's "role" (if you will) would have resulted in a "split decision" – with the geriatric psychiatrist in support of the plan and the patient's medical doctor perceiving himself as not in a position to assess the patient's ability to manage at home. Who decides?

What role, if any, is played by "informal" supporters in the absence of a "blood relative," i.e., Mr. Ansen's friend, Mrs. Stone, who had known Mr. Ansen for many years? Do the Mrs. Stones assist in the decision process?

What role is played by the older adult, him or herself? Numerous studies suggest that inducing a greater sense of personal responsibility in people who may have virtually relinquished decision making, either by choice or necessity, produces improvement; thus, some of the negative consequences of aging may be retarded, reversed, or possibly even prevented by restoring to the aged the right to make decisions.

Mr. Ansen never left the rehabilitation unit of the acute care hospital. He died the day before he was supposed to be discharged. One of Mrs. Harper's sons reconsidered his mother's wishes and brought her to his home to live; a plan to which they are adjusting remarkably well. Connecticut Community Care, Inc. remained involved with Mrs. Harper and her family to reduce the potential transition hazards so often evident in this kind of a situation.

Both of these clients did have the opportunity to leave skilled nursing facilities and return to a more familiar home setting. This is an opportunity not always achieved by many older adults in comparable situations. Who decides?

Connecticut Community Care, Inc.
Bristol, Connecticut

BIBLIOGRAPHY

1. Austin, C. D. and Seidl, F. W.: 1981, 'Validating Professional Judgement in a Home Care Agency', *Health and Social Work* **6**(1), 50–56.
2. Barney, J. L.: 1977, 'The Prerogative of Choice in Long Term Care', *The Gerontologist* **17**(4), 309–314.
3. Brody, D. S.: 1980, 'The Patient's Role in Clinical Decision-Making', *Annals of Internal Medicine* **93**(5), 718–722.
4. Dervin, J., Dervin, P. and Jonsen, A. R.: 1981, 'Ethical Considerations in Elder Care', in O'Hara-Devereaux (eds.), *Elder Care*, Grune & Stratton, New York.
5. Dreher, B: 1981, 'Deciding With the Elderly', *Geriatric Nursing* **2**(2), 122–126.
6. Langer, E. J. and Rodin, J.: 1976, 'The Effects of Choice and Enhanced Personal Responsibility for the Aged: A Field Experiment in an Institutional Setting', *Journal of Personality & Social Psychology* **34**(2), 191–198.
7. Pace, W. D. and Anstett, R. E.: 1984, 'Placement Decisions for the Elderly: A Family Crisis', *The Journal of Family Practice* **18**(1), 31–46.
8. Putt, A.: 1981, 'The Hardest Decision', *Geriatric Nursing* **2**(2), 122–125.
9. Schulz, R., and Hanusa, B. H.: 1978, 'Long Term Effects of Control and Predictability-Enhancing Interventions: Findings & Ethical Issues', *Journal of Personality and Social Psychology* **36**(11), 1194–1201.
10. Shelley, F.: 1983, 'When to Intrude in Your Parents' Lives', *Life Options* **I**(1) 43–56.
11. Smallegan, M.: 1981, 'Decision Making for Nursing Home Admission: A Preliminary Study', *Journal of Gerontological Nursing* **7**(5), 280–285.

NANCY NEVELOFF DUBLER

THE DEPENDENT ELDERLY: LEGAL RIGHTS AND RESPONSIBILITIES IN AGENT CUSTODY

I. INTRODUCTION

Individuals who have reached the age of majority are permitted a wide range of choice in American society. They may choose where or whether to work, with whom to associate, and how to pattern and place their lives. These decisions may be wise or foolish; they may enhance the quality of life or put existence itself at risk. These individual decisions and actions may encroach upon the lives of others, in violation of civil or criminal law, and thus be subject to fine or punishment. They may fall within the standard for civil commitment – exhibiting a danger to self or others – and invite involuntary restraint. Absent such circumstances, however, individual choice is largely unregulated and unsupervised.

Dependent elderly persons are often denied these rights to choose. Because they often cannot effectuate preference without assistance, their rights to choose require the cooperation of others, both individuals and agencies, thus permitting strangers to scrutinize prospective plans. This involvement of helpers and facilitators may mean that differing standards of judgment and measures of worth will be applied to an elderly individual's choice. Conflicting value systems, which often reflect competing concerns of institutional and individual self-protection and convenience, may be at odds with the elderly person's preference. Elderly dependent persons are therefore at great risk of losing their right to decide about the course and conduct of their lives.

Exploration and examination of articulable and pressing societal problems often proceed slowly. Issues must first be identified, data gathered and anecdotes arranged and abstracted to generate principles; competing principles must then be tested against the social and economic realities of contemporary society. This process of formulating attitudes, presenting rules and structuring responsibilities is presently underway in regard to the care of the dependent elderly, especially those whose articulated preference may, in the opinion of care-givers, create life-imperilling situations.([2], [14], [32])

Stuart F. Spicker, Stanley R. Ingman, and Ian R. Lawson (eds.),
Ethical Dimensions of Geriatric Care, 137–159.
© *1987 by D. Reidel Publishing Company.*

Some of the responsibilities of agencies in regard to dependent elderly are dictated by the ethics of caring; a few are required by law.

The first responsibility of caregivers is to discover and document individual preferences and desires. The second is to help to effectuate plans designed to maximize autonomy within the least restrictive environment ([10], [45], [49]). The third is to recognize and limit the paternalism of us, the arrogant young (or younger). The fourth is to create specific procedures to help maximize individual preference for persons of diminished or declining cognitive capacity. The fifth is to create procedures and guidelines to ensure that in those cases where autonomy is no longer a possible governing principle, the "best interest" of the person is identified and pursued. Finally there is the responsibility to fight against the extension of "defensive medicine" into decisions about the care of elderly persons. Fears of institutional liability must not be permitted to trump the preference of the elderly even when their voice is quiet, and their ability to effectuate decisions limited.

This discussion is limited to issues involving dependent elderly persons, i.e., those who are easily influenced and controlled by others and who must rely on others for support, and even existence [38]. The term "elderly" is clearly not a synonym for dependent. The iconography of modern American society, however, tends to support this sloppy allusion. Forcefulness can be consistent with advanced chronological age, as the Gray Panthers and other active and aggressive groups of older and retired persons consistently demonstrate. Despite such efforts, however, advertising campaigns, popular culture, and visible patterns of relationship reinforce the stereotype of elderly dependence. More critically, providers and agencies who serve the elderly not only respond to this national myth but, in large measure, help to nurture and sustain it.

The first section of this comment will attempt to expand on the concepts of agency responsibility for dependent elderly by examining briefly two elements of personhood which are generally associated with independence: (1) the concept of autonomy, as related to discussions of informed consent in the health care context (dependent elderly are most often enmeshed in health care or caring systems) ([12], [22], [24], [33]), and (2) the issue of competence, which constitutes a necessary prerequisite for informed consent ([13], [15], [17]).

The next section will focus on the case examples of dependent elderly persons in conflict with a supervising agency's perspective and power. Finally, the paper will conclude with suggestions for concrete procedures designed to implement the principles discussed at the outset. The goal of this comment is to translate ethical precepts into operating principles which can, in turn, be recognized as legally enforceable rights.

II. AUTONOMY AND COMPETENCE

It is now settled law and virtually unchallenged ethics that adult individuals, of sound mind, i.e., those who are "competent," have the right to consent to or to refuse suggested medical treatments and interventions ([43], [44], [53]). By extension, competent adults should also be able to refuse "caring," not only care. They should be able to refuse housekeeping assistance, visiting nurse supervision, or other proferred community interventions.

There is a presumption [4] in our society that once individuals reach majority, that magic moment of adulthood, they have the individual capability (independent of acquired wisdom and education) to confront various choices and decide in ways that maximize self-interest. This presumption is unquestionably contradicted in individual cases in which foolishness, self-destructiveness, or the inability to identify the objectively best choice among options results in poor decisions. Nonetheless, our societal schema dictates that empowerment in general flows from generic judgments rather than individual assessments. To require individual empowerment on a case-by-case basis would be to court chaos, discrimination, and manipulation by power elites.

This general presumption of ability certainly permits the less able to negotiate objectively foolish purchases and contracts. For example, one could purchase a used car without ever examining or testing it. It is neither in the interest of a vendor to question capability, nor in the interest of society to mandate an individual showing of capacity. To facilitate and protect important societal interests in orderly commercial transactions, we permit individuals who have reached the age of majority to execute such binding and enforceable (absent fraud) contracts. Thus, the presumption of legal competence is rarely contested in matters of commerce.

In the area of health care, the presumption of legal competence is important (although not dispositive). Physicians and health care providers are not used car salesmen. Although their object is to provide a product, i.e., to deliver service, they are constrained by the ethics of medicine and trained to identify and pursue the "best interest" of the patient, as life and health are at stake ([30], [31]). This process requires an individualized investigation. Thus, a patient's presumed competence which, outside of health care, is unquestioned, is far more likely to be the object of focussed inquiry.

The differential requirements which govern relationships on the street and regulate those between health care providers and patients provide the basis for a radical disjunction between *theories of choosing* – which empower the individual – and the *reality of choice* – in which individual voices are often unheard and individual preference unheeded. This chasm between theory and fact is produced by a number of factors.

Caregivers are trained to decide and to intervene, not to negotiate, mediate and withhold. A choosing process requires time for initiation of a relationship, the development of trust, and the sharing of specific information. All of these parts of an optimally continuous process are often cut short by episodic patterns of care and by the financial disincentives attached to communicating; discussions comprise time segments which cannot be submitted for adequate reimbursements according to prearranged categories. The abilities of a person to choose may be compromised by the institutional setting, which can be intimidating and confusing. Finally, the disparity in power between the patient or institutional resident and provider discourages real discourse. Especially in a chronic or long-term care facility, the fear of alienating the powerful and thus suffering disapproval, approbation, anger, or neglect looms large – if not as a reality, at least as a fear.

Despite these complexities, the law provides that persons who are "competent" and who are "human beings of adult years and sound mind," shall have a "right to determine" what shall be done with their own bodies [53]. Decisions regarding a body clearly encompass how it will be treated and where it will be placed.

The law is clear regarding the rights of competent adults. The law also addresses the issue of the persons who are totally and clearly incompetent to make decisions. These persons may be comatose, in a permanent vegetative state, totally demented, or severely congenitally retarded ([47], [48], [50], [51], [54]). Many state courts have now determined that there are legal principles which should guide decision-making for these persons ([47], [48], [50], [51], [54]). Some states have fashioned a doctrine of substituted judgment ([47], [54]). Others rely on the constitutional right of privacy or the common law right of self-determination

which can be exercised by a proxy [51]. Many jurisdictions have specified procedures to insure that substantive principles developed to
protect the rights of the clearly incompetent are applied rigorously [4].
Cases require the adversary presentation of facts. All procedures envision the possibility of appellate review. This panoply of legal protections
serves to preserve and protect individual interests and rights when the
individual is "incompetent." The concept of competence is key. If
present, it supports individual action; if clearly absent, it triggers alternative legal processes ([4], [13], [15]).

Competence is a legal presumption. It is one of the indicia of adulthood. Our legal system presumes competence at a stipulated age. That
age was 21 until the passage of the constitutional amendment which
lowered the voting age to 18. Subsequently, most states legislated 18 as
the age of legal majority or presumed competence [27]. This is but the
most recent fluctuation in a centuries-old pattern of readjustment.

In medieval times peasants and landed gentry reached majority or
were presumed competent at different ages. For peasantry, where
physical ability was paramount and reasoning ability less critical, the age
of majority was earlier. For the nobles, as the weight of armor increased, the age of adulthood rose commensurately [27]. The concept of
competence was then, as it is now, responsive to societal needs to
empower or limit the rights of certain sub-groups within it [27]. As such,
competence is an artifact of society – an artificial construct designed to
meet specific needs.

Competence also reflects certain judgments and perceptions about
the human condition. For example, it incorporates our collective understanding that small children and infants have neither the skill, wisdom,
experience nor intellect to make weighty decisions. They lack the
abstract ability to project consequences and to accept present burden
for future benefit. Children are thus categorically incompetent ([20],
[27]). This judgment is neither individual nor artificial but an actual
response to perceived patterns of human development.

There are thus different understandings of competence. The word,
"competence," refers to a melange of societal judgments: it both
encompasses the unchanging perceptions and psychological realities
about human development, and it reflects fluctuating societal values
which are based on, and constrained by, needs of the time (for example,
by patterns of employment needs for conscription or by the requirements of technology). As a societal artifact, competence may reflect the

timebound needs of society to include or exclude persons from full participation in the community.

This multifaceted concept of competence is central to a discussion of the rights of dependent elderly. Despite the definitions of dependence, which describe life situations and do not focus on cognitive function, decision-making capability, or intellectual acuity, there is a natural tendency for caregivers to assume that the fact of dependence compromises intellect. There may even be some basis for this assumption.

The fact of dependence may gradually erode capacities for independence and self-care which we generally associate with competent persons, and which may form some part of capable decision-making. Nursing home patients may acquire "learned helplessness." As they are asked to assume fewer responsibilities, their ability to perform independently diminishes [7]. The operation of a "total institution" [16] has been shown to diminish previously developed skills and abilities. Therefore the fact of dependence may relate inversely over time to issues of competence.

Not only are the content of the term and its definition multifaceted, but the legal settings in which it is used and the public policy or value purposes that it serves mandate different definitions for different purposes. The term, "competence," is used with widely different meanings in the law. One can be competent to stand trial, to make a contract, or to execute a will. In each setting the term varies with the public policy purpose which the construct is designed to advance. Thus in order to be competent to stand trial one must both understand the specific charges and also the process of the criminal justice system. The object is to ensure that our sense of fairness will not be offended by a trial which convicts the uncomprehending. Competence to contract, as previously noted, depends not on intelligence, wisdom, or judgment but on the ability to have a "meeting of the minds." The value inherent in that standard is support for the orderly process of commerce. Testamentary competence is again of a lesser sort. If one knows the property to be transferred, and can identify the beneficiaries of one's action, then, despite substantial intellectual decline and cognitive deficit, an executed document is likely to withstand challenge. The value inherent in that concept of testamentary capacity is the support for the orderly and regular transfer of title and property and the protection of the testator's choice of beneficiaries [17].

Most commentators agree that it takes less mental capacity to execute a valid will than it does to execute a valid contract [17]. One textbook on wills has commented, "if the testator can do this (hold in memory the natural objects of his bounty, his property, and the scope and bearing of his will) he has mind enough . . . ." [17]. Under an analysis which sees public policy as a determining factor in setting legal standards, the rules governing capacity to contract ensure the security of transactions and the security of acquisitions [17]. The rules governing testamentary capacity protect the social institution of the family by supporting the statutory preference for orderly testamentary disposition. So too, the rules for competence in health care decision-making support the agenda of the medical profession, often in the face of patient's refusal.

The health care professions often discuss the issues of competence to make health care decisions and life choices. In that context it has been suggested that in order to be competent to decide on a medical decision, one must have the ability to understand the diagnosis, prognosis, alternative treatments, the risks and benefits of those alternative treatments, and the consequences of non-intervention ([4], [12], [24], [33], [43]). The values inherent in and underlying this quite obviously rigorous policy support – in theory – the concept of self-determination and protections for the integrity of one's body.

Paradoxically, however, this rigorous litany often supports the efficient and unchallenged application of medical knowledge and skill. This is so precisely because the standard is so lofty, because these requirements for the exercise of rights are so demanding and because it is rarely in the interest of caregivers to question competence so long as the patient agrees with the care option presented. Thus the standard becomes one which is used as the yardstick of patient ability in the main, and is thus employed only when the person refuses [4]. As such it becomes the vehicle for determining non-competence, and for disempowering patient refusals. By articulating and stipulating a complex set of skills, failures to meet these tests in a contest or conflict with caregivers will support the choice of the professionals.

Analyses of the term "competence" tend to mix normative and descriptive language. In discussing competency to consent to research, one article argues that empirical analysis can rank four commonly used standards which can be arranged in a hierarchy of increasing ability. The utilization of a particular standard, however, cannot be made solely by psychiatric assessment, but requires consideration of the policy goals that use of each standard would help to advance [5].

Another related comment identifies tests of competency and finds that "the circumstances in which competency becomes an issue determine which elements of which test are stressed and which are underplayed" [33]. The tests isolated are: (1) evidencing a choice; (2) "reasonable" outcome of choice; (3) choice based on "rational" reasons; (4) ability to understand, and (5) actual understanding.

To comment on this last schema and to support the proposition that competence is a societal artifact, consider the first level of competence, i.e., evidencing a choice, a very low level of ability. It is probably adequate for deciding whether or not to wear green or blue sneakers. It was, for example, adequate for my son, when he was three years old, to decide to wear seven undershirts to school. Each one had a different picture, a Mickey Mouse, a Donald Duck, and so forth. I permitted him to wear his seven shirts even though I assumed he would be too warm; at the same time I clearly would not have allowed him at his tender age to cross the street by himself. I use this real, although frivolous, example to support the argument that competence is a combination of many factors: the abilities of the individual; the purposes, or self-interests of others and the value judgments of the more powerful (individual or society), as to the possible risks and consequences of empowerment. As such, it is an artifact created by a combination of societal needs and individual judgments in a particular context. My son was competent to make his quite bizarre decision because I granted him this liberty. I did so, moreover, because without his seven undershirts he was not going to school; since I was going to work, I needed him to be in school. His choice suited the purpose of the more powerful.

To return to the schema of five-tiered competence [33], the next two levels are again objective judgments of the powerful in light of their perceptions and needs. The last two are individual determinations, which may be able to be judged according to some measuring and objective instruments. Finally, all of these abstractions are tools for supporting or contesting statements of individual will, preference, or desire. They are not static definitions but useful tools to support or override individual decisions.

More recently, Virginia Abernathy [1] has suggested that the proper approach for understanding competence is not to further refine the positive standard, but rather to suggest a standard for determining that a patient is *not* competent to refuse treatment. She suggests that a standard to override competence should be based upon a judgment of generalized incompetence, including clear evidence that a patient is uninformable on emotionally neutral issues and cognitively incapable of making ordinary decisions on matters unrelated to the particular crisis [1].

This last suggested standard raises the distinction, critical in health care and central to our discussion of the dependent elderly, between competence to consent to or to agree with the plan, program, or intervention suggested and competence to refuse. In theory, equal capabilities should support both consent and refusal; but in fact they don't – and perhaps they should not. Practically, it is in no one's interest to contest the capability of a patient, who consents to a plan presented. Caregivers are, however, appropriately concerned with a patient's decision which appears to contravene the collected skills and wisdom of helping professions. Refusals tear apart the fabric of care. Consent adds one more delicate thread to that fabric. Care giver acquiescence in consent or refusal which does not reflect considered judgment and real individual value determinations may not serve the needs of the patient and may in fact leave that patient abandoned – unable to identify self-interest, with no advocate arguing for objective best interest.

Competence is thus a societal artifact and a facet of power relationships. It is also decision-specific. That is, a person can be competent to decide between choices of little consequence (e.g., undershirts) and not competent to decide in a complex situation with great risk attached to one course of action ([18], [23]).

In a work which has come to be regarded as reflecting moral consensus, the President's Commission for the Study of Ethical Problems in Medicine and Biomedical and Behavioral Research viewed competence as consisting of three elements:

(1) possession of a set of values and goals;
(2) the ability to communicate and to understand information; and
(3) the ability to reason and to deliberate about one's choices [31].

The Commission recognized, despite the rigor of its definition, that emotional states, social settings, and life experience play a role in competence. Nonetheless, its definition contrasts markedly with the previous definitions, albeit in different contexts, in which "holding in memory," or "mind enough" are sufficient to empower individuals [17].

Emerging from the discussion above are two distinct but largely overlapping senses of competence:

(1) It is a synonym for what society will allow people to do, in different settings, at different stages of life; and
(2) It is a set of intellectual and psychological capacities that are necessary for informed consent, i.e., which subordinate the judgment and action of others to the autonomous designs of an individual.

If someone is severely demented, she is unlikely to qualify under the second, yet may still be required to be treated "as if" competent, . . . because it suits the purposes of the powerful. In such circumstances the less capable or incompetent patient may be truly abandoned. She may be unable to identify and advocate her self-interest and cannot rely on her best interest being the concern of deciders.

It does violence to language and common sense to insist that the cognitively impaired person is "competent" to make decisions. The reason we do so, however, is to ensure that the patient's preference, when previously capable, is respected. A preference for living at home, for example, may remain strong and constant, even with diminished cognition.

What is needed is a new legal doctrine. Instead of the "substituted judgment" doctrine ([47], [54]) – what would this person want if she could tell us – we should employ the doctrine of "supported judgment," – what is this person's "spoken choice" [39] – what is her present articulated preference and how does it relate to prior patterns of preference. If such a substantive doctrine existed and was operative, then the decision maker would be required to discover and account for the patient's previously competent wishes and to take into account the psychic costs of overriding a statement and forcing treatment on an unwilling patient.

A determined desire can probably not reflect autonomy in a very severely impaired patient. It may well, however, be the expression of prior strong preference. As such it should not be overridden lightly.

There are many people with diminished capacity and indeed quite severe degrees of dementia who, if challenged, would not meet the more demanding standard of the President's Commission. They might not have sufficient mental ability to understand in depth the choice being made and, more importantly, might lack the ability to communicate a decision [25]. On the other hand, if there is no one to challenge and to question, competence will be assumed and patient ability to choose will be preserved.

III. CONSERVATORSHIP AND GUARDIANSHIP

If a person's competence is challenged, the law now provides a route for the resolution of the issue and for the appointment of a proxy decision-maker, should the person be found incapacitated. Since medieval times there has been the assumption that certain persons – e.g., children and the lunatic or idiot adult – need specific state intervention for their protection and for the protection of their property. These realizations led in time to development of conservatorship and guardianship statutes in all states. Although variously labeled, conservatorship usually means the appointment of someone with power over the property of another, and guardianship means the appointment of someone with power over the property and person of another ([3], [6], [8], [9], [25], [26], [29], [35]). Guardianship proceedings generally require a judicial finding of "incompetence" before appointment of a surrogate decider; conservatorships should be dependent on an examination of the functional abilities of an individual. The first makes a judgment on the *gestalt* of the person, whereas the other is supposed to judge skills.

Consider the language and the operation of the New York statute as an example of underlying and conflicting messages and of specific assumptions about dependence, skills, and aging.

The New York Conservatorship statute, Article 77 of the Mental Hygiene Law, states in part:

The Supreme Court and the County Courts outside the City of New York, if satisfied by clear and convincing proof of the need therefor, shall have the power to appoint one or more conservators of the property for a resident who has not been judicially declared and who by reason of advanced age, illness, infirmity, mental weakness, intemperance, addiction to drugs or other cause, has suffered substantial impairment of his ability to care for his property or has become unable to provide for himself or others dependent upon him for support . . . [41].

In the New York statute, it must be established, by clear and convincing proof, that the proposed conservatee has a specific infirmity (Element #1), which has led to a substantial impairment of his ability to manage his property (Element #2), which has resulted in the need for the appointment of a conservator (Element #3). These judgments could be arrived at independently. The first often requires medical testimony, the latter two, independent evaluations. In fact, medical testimony on the first often suffices to prove all elements and thus to deprive a person of the ability to exercise authority over property. In a recent amendment to the law this surrogate may also exercise some aegis over the person [42].

Guardianship statutes lead not to functional supervision, but categorical infantilization. They reduce an adult to the status of a child in the eyes of the law. By operation of these statutes, an individual is generally precluded from buying or selling property, entering into contracts, suing or being sued, marrying, operating a motor vehicle, changing his residence, writing checks or engaging in financial transactions of any kind [35]. In addition, most statutes prohibit the individual from executing a will, and most do not permit him to vote [35]. The underlying judicial judgment which permits the appointment of a guardian is one which declares an individual to be "incompetent," about which enough has been said.

Most persons subjected to the operation of guardianship and conservatorship statutes are dependent and elderly. That is hardly surprising when, as in the New York statute, advanced age is provided as one basis for statutory intervention [41].

A clear problem alluded to above is that in both guardianship and conservatorship proceedings, there is the real possibility of an *ex parte* hearing, that is, a "non-adversarial" proceeding. Not only is the elderly person absent from the proceeding, but there is, in addition, no counsel present to represent, and advocate on behalf of, the elderly individual's position. A guardian *ad litem* is appointed by the judge in all proceedings; her role, however is not to be an advocate but rather to be an independent evaluator of the "best interest" of the proposed ward [36]. The dependent and aged individual is thus, in fact, unrepresented and, in essence, ignored in a proceeding that overwhelmingly impacts on the individual's future lifestyle and independence. This reflects the societal value that the orderly management and transfer of property, and the smooth functioning of society, are paramount to the protection and perhaps preservation of an aged individual's perspective and sense of autonomy.

Since neither the process of aging nor the notion of mental debility is well understood by judges, the statute probably provides a good deal of "cuing." If the judge is in doubt as to whether a person might be unable to manage properly, he is cued that people might be unable to manage because of disease or weak mind, or by reason of old age. The legislature appears to have made a determination that, as with disease, aging may well cause inability to manage property. Therefore, the judge is more prepared to find a person who is physically feeble to be a bad property manager ([3], [14]).

In whose interest is it to raise these issues? In New York, for example, a study demonstrated that the single largest petitioner for guardianship was the state itself, arguably facilitating the orderly administrative process "on behalf of" the elderly and infirm ([3], [15]). The effect of these statutes is essentially to limit the abilities of the aged to reassess their positions or to act eccentrically in support of newly-defined goals.

These statutes can be used properly and benevolently to protect and support the feeble, the lonely, and the confused – the sad detritus of society, for whom some legal protective status must be established before help can be provided.

Both conservatorship and guardianship statutes can, in fact, be used fairly and equitably. This can only be the case, however, if the aged person is present, is represented by counsel to argue her position, and is provided with a fair and impartial judgment according to the language of the statute based on functional ability, not on hearsay or conclusory medical judgments.

IV. THE SETTINGS AND THE DECISIONS

Both of the cases presented describe the. earnest efforts of elderly patients to effectuate their preference in opposition to the honest concern of health care providers to structure the safest environment possible. The cases are different in one important aspect. In the first, although the discharge plan could be seen as unwise, it was clearly devised by a man of unquestioned intellectual and cognitive ability. The clash was clearly one of value preference. Health care providers, aware of his physical deficits, would have preferred him to be in a caring and capable medical environment, able to provide the "best" care. In Mr. A's hierarchy of values, however, being cared for was less important than being in a home with which his life and history were intertwined. He had, I would argue, the clear unequivocal right to make that decision and to have that decision respected. The responsibility of all parties was then to act to assist him, given his physical frailty and lack of indepen-dent ability to effectuate his decision.

Mrs. S presents the more usual sort of problematic case. Whereas Mrs. S wanted to return to home, she had both cognitive and intellectual deficits and some impaired judgment. Nonetheless, her spoken choice or articulated preference was clear. Moreover, despite demonstrated deficits, this choice, it could be argued, was based on a consistency of theme and a history of a pattern of preference. This pattern had two elements: first, a clear attachment to her home, and second, a strong preference for dying in it rather than in an institution. These wishes were articulated in light of the history of the lives and deaths of loved ones. Mrs. S wanted to die at home. This "knowing" of one's mind existed despite the documented intellectual losses. It too, I would argue, deserved respect and all efforts of support. A "supported judgment" would have indicated respect for her wishes.

I have come to call self-awareness of strong preference, which survives cognitive decline and dementia, a "sedimented life preference" – that is, a consistent theme built up by actions, thought, and behavior over time in the context of personal history. These themes are so strong and so fundamental to a human soul that they survive intellectual and physical decline. When this sort of theme emerges, buttressed by surprisingly articulate statements (such as those describing the death of loved ones at home) and supported by confirmatory statements of family and friends, it deserves great weight and respect; in some cases, such as that of Mrs. S, it should be permitted to override issues of present incapacity, if this resolution is even marginally possible. The responsibility of caregivers should be to arrange for support for this preference over more abstract notions of "best interest."

These cases pit the right of a physically or intellectually compromised person to make decisions which stand opposed to the wisdom of others. The questions they presented are: first, when should these expressions of preference be respected or overridden? Second, what criteria and standards could be suggested to guide the decision? Third, could provider action, in support of the individual, deviate from accepted standards of "best interest" and subject the individual or institution to possible liability?

Elderly persons are not like children for whom a judgment of "best interest" is always appropriate. Elderly persons have developed skills, desires, likes and dislikes, and complex emotional reactions which in combination may counterbalance and outweigh generalized notions of "best interest." Although the process of illness may have eroded certain abilities, the nugget of values assumed earlier in life may still be intact. Despite obvious deficits, there may be this residual autonomy – a track record of preference. This constancy of theme which survives dementia and debility, this "sedimented life preference," should be supported when buttressed by independent evidence, prior statements and contemporaneous utterance, even if small and weak. This respect, of course, does not require slavish and uncritical obeisance. In the cases presented, the elderly person was quite clear, despite deficits, that he or she did not want the solution presented by a caregiver or responsible agent. I would argue that the primary responsibility of the agency is to respect these statements, subject to the procedures outlined in the conclusion.

V. ROUTES FOR THE RESOLUTION OF CONFLICT (CRITERIA AND STANDARDS)

The conflict presented by the case examples is a conflict between the right of a patient with compromised abilities to decide and the obligation of the involved agency to protect that person or itself from the potential negative effects of the decision. Possible routes for resolution include (singly or in combination):

1. the use of existing conservatorship and guardianship statutes,
2. the extension of that model of intervention to community guardian or protective services agencies,
3. the further development of elder abuse reporting laws which apply to psychological abuse or inappropriate restraint,
4. the development of an elder court – a new jurisdictional solution, and
5. the creation of regularized efforts to document prior preference for the support of later decisions.

Conservatorship and guardianship actions have been mentioned previously. There is a new development, however, which is the extension of this difficult, cumbersome individual action to public adult protective services. In 1977 the 95th Congress produced a working paper, 'Protective Services for the Elderly' [37]. The model Adult Protective Services Act based on that working paper authorizes "the imposition of protective services through guardianship for an elderly person who lacks the capacity to consent to receive protective services" [21]. The object of the act is to provide a less cumbersome mechanism at that point when an uncooperative elderly person reaches active resistance [25].

The history thus far in those states which have substantial experience with a variation of the law is that the public guardian and its office are a "prime mover" in the institutionalization of wards with involuntary commitment repeatedly constituting the only "service" provided [25].

Although in theory the office of the public guardian is intended to deliver service enabling elderly persons to remain in the home, in practice they have emerged as agencies of social management able to move the poor and the near poor out of the community [25]. In fact, these agencies have been a major force in removing the elderly from their homes against their will.

A bill recently introduced in the New York State legislature to set up a community guardian program attempts to meet this criticism by stipulating that the purpose of the guardian is to manage the personal and financial affairs of the individual to enable him to continue to live in the community [40].

The object of the guardian, however, is not to search for fashion or cajole consent but rather to offer consent over the refusal of the individual.

There is a proper although limited purpose for such bills. Under the present court system, people who are functionally unable to manage funds, who are not refusing care but who need financial support and organization, i.e., the appointment of a conservator, will find that the conservator is unavailable unless the assets to be managed are substantial. Conservators, political appointees of the court, are generally unwilling to assume obligation for persons whose assets will not provide sufficient remuneration for their efforts. The community guardian, as financial facilitator for an elderly person receiving minimal social security and pension, would indeed be a welcome addition. Historically, however, the community guardian has often acted to provide the least resistant alternative for him or herself rather than the least restrictive environment for the ward [3]. The least trouble for the community manager comes from the quickest institutionalization.

There is substantial case law dealing with the right to deinstitutionalization ([45], [49]), the right to treatment ([46], [52]), and the development of community facilities. The thrust of the decisions mandates that the "least restrictive alternative" is the most appropriate setting for the less competent [49]. The least restrictive alternative for most dependent elderly is clearly the home.

The danger of these community guardian efforts may lie in the fact that the protective services legislation merely "creates a separate agency and provides legitimization for coercive state intervention which heretofore was employed either without explicit legal authority or through (more cumbersome) guardianship and civil commitment laws" [25].

In whose interest is a public guardian program? There are certainly some elderly persons for whom ancillary financial management services will secure and maintain them in the home as per their preference. The office of the guardian, however, may provide a ready site focus for complaints. Neighbors may disapprove of the deteriorated state or apartment of a tenant. Landlords may see it as a ready route for easy eviction. Health care providers and social service agencies all have ready reference to easy support for their understanding of what is in the best interest of the person. The danger is that the public guardian will become an agency of social management permitting the easy coercion of the frail [28].

The extension of the model of child abuse reporting legislation to the elderly has also been suggested as a way of helping to uncover elderly at risk for neglect and want. Again, the disanalogy with child abuse reporting statutes is evident. There is an *a priori* responsibility on the part of the state to offer protection to children as categorically dependent persons, if their natural protectors, i.e., parents, fail in their responsibilities [34]. There is no comparable obligation of the state to assume the incompetence of the elderly person and therefore his need for protection. Analysis of situations of elder abuse (which never mention the abuse of inappropriate institutional confinement) indicates that the continuing presence of the abusing situation may reflect the victim's determination that "it is better to stay in a situation that is less satisfactory than suffer the consequences of professional intervention" [20]. Complex patterns of family interaction in which neglect and abuse are found may reflect decades of dysfunctional family behavior. Despite that, the benefit from continued connection may outweigh the alternative of sterile care by strangers.

The most troubling aspect of elder abuse reporting laws reflects the criticism of elder protective services. It provides the "potential for state intervention in the lives of those who want to be left alone" [20]. It would be virtually impossible and probably a bad idea to extend the concept of abuse to situations of possibly inappropriate housing of elderly persons.

The experience with other protective services for the elderly has shown that many things which are done under the guise of protection actually deprive the individual of dignity, autonomy, and self-esteem. In addition, it should be noted that one of the major disabilities of the child abuse reporting laws, where the premise of need is uncontroversial, has been the inability of society to provide sufficient agency support to respond effectively to reporting [20]. The quality of the staff responding to complaints, the adequacy of the funding, and the remedies available in response for most child abuse reporting situations are not calculated to improve the lot of many children – the same would more than likely be true in regard to the elderly.

The problem with all of the alternatives discussed above is the easy route they provide for the imposition of caregivers' will over the preferences of the elderly. Part of this problem could theoretically be avoided by the creation of a whole new court system comparable to the family court which could serve for the adjudication of elder issues. By recognizing the problem of dependent elderly with diminished capacity, this jurisdictional solution would create a forum for these people to present their views with the full support of advocates in a rigorous adversarial surrounding. Again, in theory, this might provide maximal support for residual autonomy; in fact, it would most likely fall prey to the problems that beset presently existing family courts and that have attached to elder protective services legislation. That is, it would develop overly favorable assumptions about professional judgments and a bias against the "defendant or elder person." It would probably take on the quasi-judicial, paternalistic perspective that plagues the juvenile court system. The services would probably be inadequate: it would be a court for the poor. And it would soon come to share in the second class status which lack of money and position signal in our society.

The thematic connection in both of the cases presented is the inability of caregivers and society to accept that decisions which may diminish the quality or shorten the duration of life can reflect genuine desire rather than dementia. Previously documented expressions of preference should be accorded great weight. In health care, there is the growing recognition of the need to elicit patient values early in the physician-patient relationship before the patient is confronted with an acute episode of illness or with a dementia. Durable powers of attorney, living wills, the appointment of agents for health care decisions all permit competent patients to document their preferences, should they be rendered incompetent by trauma or by advancing disease [30], [31]. These prior statements are always subject to attack on the grounds of lack of present applicability caused by changed conditions or intimations of mortality. If not revoked, however, they can provide that independent evidence to support idiosyncrasy, to amplify and support a "sedimented life preference" and to encourage caregivers to act on individual desire over generalized, abstract notions of "best interest." Perhaps preferences should be elicited not only in regard to "heroic" care, but also in regard to life in "long-term care."

At stake are comfort and suffering. We, as a society, do not want our elderly persons to suffer. Nor do we want their neighbors to be inconvenienced or imperilled. Nor do we want caregivers to be faced with situations that compromise their professional ethics and/or court liability. These may, in some cases, be mutually exclusive desires.

These cases confront caregivers committed to humane care with uncomfortable options. On the one hand, they can follow clearly competent statements or compromised but firm opinions opting for discharge plans that will satisfy the patient but may contravene accepted practice. On the other, they may restrict freedom and hold a frail person against her will. By so posing the opposition it seems clear that the first route is preferable. Could it, however, subject the individual caregiver or institution to liability? If the family is in agreement, if risks are identified, if a reasonably safe (if not perfect) plan is devised, and if the patient acknowledges the risks, liability becomes ever more remote, although never entirely ruled out.

In the case of a patient who has consistently refused "caring" where no private resolution is possible, court referral before confinement is a necessity. Only a court can override a strong and consistent patient refusal of institutional care, even if the patient is cognitively impaired or somewhat demented. In the allocation of authority to professional sub-divisions of society, physicians have been charged to "care" and "do no harm," judges and courts are mandated to protect rights, divest people of authority in certain circumstances, protect the less capable from fraud and duress, and adjudicate controversy. An elderly person's refusal of institutional care is thus properly a subject for judicial scrutiny.

There is also the weighty burden that caregivers may be reasonably certain that, given the inadequate representation for proposed conservatees, the providers' presentation will win. The court becomes a sanction for professional judgment.

Cases requiring such recourse should not and will not be the norm. Courts are not equipped procedurally, financially, and temperamentally to deal with huge numbers of treatment decisions. Experience indicates that the number actually presented to court will be miniscule. Some refusals will gain strength from support of family, friends, and clergy, causing providers to rethink the necessity for judicial review. Many will change with time, explanation, and exhortation, and perhaps subtle coercion.

VI. RECOMMENDATIONS AND CONCLUSIONS

These case studies presented by Kataja and Gavin illustrate the need for interdisciplinary consultation as part of the appropriate treatment for certain types of dependent geriatric patients. Comprehensive geriatric care for dependent persons requires the pooling of information.

A possible procedure could include:

1. A conference of all participating disciplines convened
   a. to evaluate the discharge options,
   b. to determine the best course of placement in the context of the patient's documented mental and physical abilities and disabilities,
   c. to consider how these disabilities affect perceptions and skills, and
   d. to weigh possible outcomes of the various placement and discharge decisions in terms of the likelihood of harm and the possible immediacy of harm.
2. The contemporaneous declarations of the patient should be ascertained and an attempt made to determine if these statements reflect autonomy, "a sedimented life preference" or if they are reflexive verbal responses to the situation. Previously executed documents and prior discussions and/or decision making are essential in this respect.
3. If the conferees conclude that the patient's cognitive ability and judgment are compromised, and that her stated preference is clearly not in her best interest, suit should be instituted in a court of

appropriate jurisdiction alleging that agency support for the elderly person's stated preference could lead to immediate and foreseeable harm.

4. This final step – court intervention – will only be relevant in a tiny number of cases. A patient consenting to the proposed care will be seen by the staff as acting in her own best interests and thus, care will proceed. An adamant refusal should trigger judicial review not only to protect the rights of the patient but to support the integrity of the agency.

These rules will not dispose of all cases. They should provide some guidance for the frustrated, angry, and often appropriate response of caregivers determined to do what "is best," since what is "best" for a dependent elderly person is precisely what is at issue.

Montefiore Medical Center
Bronx, New York

BIBLIOGRAPHY

1. Abernethy, V.: 1984, 'Compassion, Control and Decisions About Competency', *American Journal of Psychiatry* **141**, 53.
2. Alexander, G.: 1980, 'Remaining Responsible: On Control of One's Health Needs in Aging', *Santa Clara Law Review* **20**, 13.
3. Alexander, G.: 1977, 'Aging in America – IV; Who Benefits from Conservatorship?', *Trial* (May) 30.
4. Annas, G. and Glantz, L.: 1985, 'Withholding and Withdrawing of Life-Sustaining Treatment for Elderly Incompetent Patients: A Review of Court Decisions and Legislative Approaches', prepared for Office of Technology Assessment, Congress of the United States.
5. Appelbaum, P. and Roth, L.: 1981, 'Clinical Issues in the Assessment of Competency', *American Journal of Psychiatry* **138**, 1463.
6. Atkinson, G.: 1980, 'Towards a Due Process Perspective in Conservatorship Proceedings for the Aged', *Journal of Family Law* **18**, 819.
7. Avorn, J.: 1982, 'Induced Disability in Nursing Home Patients: A Controlled Trial', *Journal of the American Geriatrics Society* **30**, 397.
8. Bell, W., Schmidt, W., and Miller, K.: 1981, 'Public Guardianship and the Elderly: Findings from a National Study', *The Gerontologist* **21**, 194.
9. Campe, C.: 1985, 'Conservatorships: Their Application to the Hospitalized Elderly in New York', unpublished paper.
10. Cohen, E.: 1985, 'Caring for the Mentally Ill Elderly Without De Facto Commitments to Nursing Homes: The Right to the Least Restrictive Environment', Congressional Office of Technology Assessment, Workshop on Surrogate Decisionmaking.

158   NANCY NEVELOFF DUBLER

11. Childress, J.: 1982, *Who Should Decide? Paternalism in Health Care*, Oxford University Press, New York.
12. Drane, J.: 1984, 'Competency to Give Informed Consent: A Model for Making Clinical Assessments', *Journal of the American Medical Association* **252**, 925.
13. Drane, J.: 1985, 'The Many Faces of Competency', *Hastings Center Report* **15**, 17.
14. Dubler, N.: 1981, 'Assumptions in the Law About the Process of Aging', unpublished paper.
15. Freedman, B.: 1981, 'Competence, Marginal and Otherwise', *International Journal of Law and Psychiatry* **4**, 52.
16. Goffman, I.: 1961, *Asylums*, Anchor Books, New York.
17. Green, M.: 1940, 'Public Policies Underlying the Law of Mental Incompetency', *Michigan Law Review* **38**, 1189.
18. Hamerman, D., Dubler, N., Kennedy, G., and Masdeau, J.: 1986, 'Decision Making in Response to an Elderly Woman With Dementia Who Refused Surgical Repair of Her Fractured Hip', *Journal of the American Geriatrics Society* **34**, 234.
19. Holder, A.: 1985, *Legal Issues in Pediatrics and Adolescent Medicine*, Yale University Press, New Haven.
20. Katz, K.: 1980, 'Elder Abuse', *Journal of Family Law* **18**, 695.
21. *Legislative Approaches to the Problems of the Elderly: A Handbook of Model State Statutes*: 1978, Sponsored by The National Council of Senior Citizens, Published by Legal Research and Services for the Elderly, Washington, D.C.:
22. Meisel, A. and Roth, L. 1983, 'Toward an Informed Discussion of Informed Consent: A Review and Critique of the Empirical Studies', *Arizona Law Review* **25**, 265.
23. Melnick, V., Dubler, N., Weisbard, A., and Butler, R.: 1984, 'Clinical Research in Senile Dementia of the Alzheimer Type: Suggested Guidelines Addressing the Legal and Ethical Issues', *Journal of the American Geriatrics Society* **32**, 531.
24. Miller, L.: 1980, 'Informed Consent: I, II, III, IV', *Journal of the American Medical Association* **244**, 2100.
25. Mitchell, A.: 1978, 'Involuntary Guardianship for Incompetents: A Strategy for Legal Services Advocates', *Clearinghouse Review* (December 1977), 451.
26. Mitchell, A.: 'The Objects of Our Wisdom and Our Coercion: Involuntary Guardianship for Incompetents', *Southern California Law Review* **52**, 1405.
27. National Commission for the Protection of Human Subjects of Biomedical and Behavioral Research: 1977, 'Appendix to Report and Recommendations: Research Involving Children', Department of Health, Education and Welfare Publication No. (05) 77–0005.
28. O'Malley, T., Everitt, D., O'Malley, H., and Campion, E.: 1983, 'Identifying and Preventing Family-Mediated Abuse and Neglect of Elderly Persons', *Annals of Internal Medicine* **98**, 998.
29. Peters, R., Schmidt, W., and Miller, K.: 1985, 'Guardianship of the Elderly in Tallahassee, Florida', *The Gerontologist* **25**, 532.
30. President's Commission for the Study of Ethical Problems in Medicine and Biomedical and Behavioral Research: 1983, *Deciding to Forego Life-Sustaining Treatment*, Concern for Dying edition, New York.
31. President's Commission for the Study of Ethical Problems in Medicine and Biomedical and Behavioral Research: 1982, *Making Health Care Decisions*, Concern for Dying edition, New York.

32. Regan, J.: 1981, 'Protecting the Elderly: The New Paternalism', *Hastings Law Journal* **32**, 111.
33. Roth, L. *et al.*: 1977, 'Tests of Competency to Consent to Treatment', *American Journal of Psychiatry* **134**, 279.
34. Schwartz, A. and Hirsh, H.: 1984 'Child Abuse and Neglect: A Survey of the Law' in Carmi, A., Zimrin, H. (eds.), *Child Abuse*, Springer Verlag, Berlin.
35. Sherman, R.: 1980, 'Guardianship: Time for a Reassessment', *Fordham Law Review* **49**, 350.
36. Solender, E.: 1976, 'The Guardian Ad Litem: A Valuable Representative or an Illusory Safeguard?', *Texas Tech Law Review* **7**, 619.
37. United States Senate Special Committee on Aging, 95th Congress, 1st Session: 1977, 'Protective Services for the Elderly – A Working Paper'.
38. *Webster's New Collegiate Dictionary*, 1979, G. & C. Merriam, New York.
39. Zuckerman, C.: 1984, unpublished communication. "The Patient's 'Spoken Choice' is defined as an articulated preference but one which, due to the patient's cognitive impairment, it would be inaccurate to term a reasoned decision."
40. New York Law Section ____, A. 1240/S.5677 (1986) (awaiting Governor's signature).
41. New York Mental Hygiene Law, Section 77.01 (McKinney 1978).
42. New York Mental Hygiene Law, Section 77.21 (McKinney 1978).
43. *Canterbury v. Spence*, 464 F. 2d 772 (D.C. Cir. 1972).
44. *Cobbs v. Grant*, 502 P. 2d 1 (Cal. 1972).
45. *Covington v. Harris*, 419 F. 2d 617 (D.C. Cir. 1979).
46. *Donaldson v. O'Connor*, 422 U.S. 563 (1974).
47. *In the Matter of Quinlan*, 355 A. 2d 647 (N.J. 1979).
48. *In the Matter of Spring*, 405 N.E. 2d 115 (Mass. 1980).
49. *Lake v. Cameron*, 364 F. 2d 657 (D.C. Cir. 1966).
50. *Matter of Dinnerstein*, 380 N.E. 2d 134 (Mass. 1978).
51. *Matter of Storar, In the Matter of Phillip Eichner*, 52 N.Y. 2d 363 (1981).
52. *Rouse v. Cameron*, 373 F. 2d 451 (1966), affirmed on rehearing 387 F. 2d 241 (D.C. Cir. 1967).
53. *Schloendorff v. Society of New York Hospital*, 105 N.E. 92 (N.Y. 1914).
54. *Superintendent of Belchertown State School v. Saikewicz*, 370 N.E. 2d 417 (Mass. 1977).
55. *Wyatt v. Stickney*, 344 F. Supp. 387 (Ala. 1972).

MARGARET P. BATTIN

CHOOSING THE TIME TO DIE: THE ETHICS AND ECONOMICS OF SUICIDE IN OLD AGE

In recent discussions of distributive justice in health care, an ancient notion is again achieving currency: that when there are not enough resources to go around, the claims of the old – who no longer make economic contributions to society and who have already lived beyond a normal lifespan – diminish in comparison with the claims of the young. When not all needs and preferences can be satisfied, priorities in health care should go to those who remain productive and whose lives are not yet complete. This distributive issue about the relative strengths of claims of the elderly versus those of the young will be, no doubt, a central philosophical and economic issue for geriatrics in the 21st century – the topic to which this volume is addressed – and is very quickly becoming a pressing issue in the current century as well.

In what follows, I'd like to examine the implications of this issue for questions of how we want to die. Part 1 will show why suicide may appear to be an attractive solution to the risks of old age, both for the individual self-interest maximizer and for society as a whole. Part 2 will conjecture that in a cost-conscious climate, this attractiveness could engender societal expectations that the individual *ought* to choose suicide in preference to dependency in old age. Part 3 will examine whether there is any moral warrant for such expectations by considering age-rationing policies, and Part 4 will describe what forms age-rationed disenfranchisement from treatment might take. Considerations of mercy and of the "erosion" problem require, however, that disenfranchisement be practised not as denial of treatment, but as encouragement of suicide, and societally expected suicide would be neither immoral nor unjust. But this conclusion is disturbing, especially in a society where the background institutions are not just, and Part 5 will offer several ways of avoiding it.

1. THE RATIONAL ATTRACTIVENESS OF SUICIDE IN OLD AGE

When as rational persons anxious to promote our own welfare we

161

Stuart F. Spicker, Stanley R. Ingman, and Ian R. Lawson (eds.),
Ethical Dimensions of Geriatric Care, 161–189.
© 1987 *by D. Reidel Publishing Company.*

survey the prospects extreme old age can offer us, the vision may not be
an attractive one. We can expect, if we are realistic, physical debility,
loss of capacities for independent living, reduced sensory capacities,
disruption of affectional ties as spouse, friends and siblings die, greater
likelihood of chronically painful conditions like arthritis, less attractive
personal appearance, and, often, reduced financial status. At the same
time, however, our medical expenses will rise, whether borne by ourselves or by third-party insurers, and the amount of care we require
from our families or from institutional personnel will increase enormously. Our gait will begin to shuffle, our vision will dim, our hearing
will deteriorate, and even our sense of taste will grow dull. We will lose
control not just over our pocketbooks and our bladders but over all the
major circumstances of our lives, including where to live, with whom to
associate, what to do, and what to eat. We will become almost completely dependent, except perhaps in the most trivial ways, and with that
dependency lose whatever capacities for autonomous choice we may
now have. There will be no recovery from this condition, and though we
may hope for temporary periods of stabilization or improvement, by
and large our condition and circumstances will continue to deteriorate
until we eventually die.

This forecast of extreme old age may seem exaggeratedly negative,
and indeed for many very old persons conditions are very much more
kind. Senior citizens' and gray-power groups rightly denounce stereotyped
predictions such as these, warning that uncritical acceptance of them
invites the very conditions they predict. Nevertheless, this grim vision
does accurately reflect what lies ahead for some of us, and profound
late-life dependency, necessitated by the deterioration of our physical
and mental capacities, is what many of us can realistically expect.

In this situation, suicide may seem to be an attractive option. By
bringing about death before you suffer the most profound of the losses
to be expected, suicide serves as a strategy for sparing yourself the
miseries of old age. It permits you to retain a degree of autonomy: in
voluntary death, you choose the time, place, means, and circumstances
of dying, and need not wait until extended, expensive medical collapse
overtakes you. You can thus control both the experiential and financial
costs of dying: you die in your own way, at your own price. Suicide here
need not be equated with hanging yourself in the basement or blowing
your brains out on the back porch, but rather with a deliberate choice to
bring one's life to an end now, whether by taking a lethal dose of drugs
(with or without the assistance of family members, friends, or a physi-

cian), by deliberate self-starvation, or by refusal of life-prolonging medical treatment in order to die.[1] Indeed, the more supportive familial, social, and medical assistance in suicide becomes, the more feasible and attractive it may begin to seem as an option, particularly in the face of extreme old age. It is, after all, not a choice between life and death, but between death now and death later, and between death under circumstances within one's control and death where no one can do anything to help. In fact, argues Robert Kastenbaum, suicide can be expected to become the *preferred* mode of death, since it permits control over all the relevant aspects of dying ([25], pp. 425–41).

Such a view invites a number of objections. Many of these can be subsumed under the general claim that for the rational self-interest maximizer, suicide is always a prudential error, since if completed it can neither serve a person's interests nor satisfy rationally developed preferences. Because suicide brings about *death* and hence completely extinguishes a person's existence, it is always a prudentially erroneous thing for a person to do.

Replies to this general objection may take two forms, those that attend to the direct frustration of interests, and those that are concerned in addition with precluding the positive satisfaction of interests. Generally, a reply of the first form points out that, at least on a secular, no-afterlife metaphysics, suicide can never be an imprudent, bad choice, even for a rational self-interest maximizer, since if it is successful the agent does not have to live with the results ([26], pp. 148–9). Replies of the second sort point out, however, that although completed suicide does not result in frustrations of interests or harms which the agent experiences, it does preclude the satisfaction of many of his antecedent interests nevertheless. So, for instance, suicide precludes engaging in the activities one enjoys or in activities which would provide direct satisfaction of interests, including occupational, recreational, and other goal and non-goal-directed interests. Suicide interrupts one's projects, whether these involve writing a book or building a birdhouse; it precludes further sensory experience, like hearing music, tasting flavors, feeling the wind in one's face; it prevents human communication, including talking, planning, discussing, confiding, loving, and so on. These are important interests, the very stuff of human life. It is true that the person who does kill himself will not be aware that the satisfaction of these interests is now precluded; nevertheless, on this argument, the precluding of the satisfaction of these interests is a harm, and a compelling reason to reject suicide.

Although this second form of reply poses a much more difficult prudential case against suicide, it still will not entail that suicide in old age is always a prudentially erroneous choice. For although elderly people, like younger ones, enjoy activities, have projects, care about communication with their partners, family, and friends, enjoy hearing music, savoring flavors, and feeling the wind in their faces, for some persons in conditions of extreme old age and pronounced dependency many of these interests risk negative rather than positive satisfaction. For instance, although a person ordinarily has interests in the performance and completion of his projects – writing the book, building the birdhouse – in extreme old age the deterioration of intellectual and motor skills may make this impossible, or actually undermine work done earlier. Inept rewriting or bad last chapters spoil the book; an awkward blow of the hammer demolishes the birdhouse. Often, elderly people are diverted from such projects either by intervening circumstances or the paternalistic interventions of others, but the fact remains that one's interests in performing and completing projects can no longer always be satisfied. Substituted activities, like playing bingo or being read to, may not satisfy one's interests to the same degree, and in any case, as the deterioration of old age progresses, one may simply sit in a wheelchair or lie in a bed, and no longer be capable of even such minimal projects as these.

Similarly, although the satisfaction of sensory interests is a good, in extreme old age this too may be thwarted or undermined. Sensory experiences may be extremely limited in old age, and many of them are bad. Stereotypically negative views of nursing homes paint the diet as bland, the background noise as incessant, the beds as uncomfortable, and the air as tinged with the permanent odors of disinfectant and urine. Fresh breeze and the warmth of the sun are ancient myths, since many of the extremely elderly never go outdoors. Conditions for the non-institutionalized dependent elderly may be equally unpleasant, depending on the concern of family members or others providing care. Then too, possibilities for human communication may be similarly curtailed in extreme old age. Communication may still be possible as long as one is able to exchange a smile, or for that matter a grimace. But such small human activities as walking hand in hand are no longer possible when one cannot walk and one's hands are knotted closed with arthritis. Carrying on a conversation is not possible if deafness is advanced or one has lost control of one's voice or if one is intubated or on respiratory supports. Making love, even if one were to find privacy and a partner, may be perceived as a cruel joke. In any case, one's closest affectional partners are likely to be already dead, institutionalized, or removed from day-to-day contact by external circumstances and their own deterioration. As memory fails, especially in conditions like Alzheimer's, a person may experience severe disorientation and confusion, and ultimately may no longer recognize affectional partners at all. Indeed, communication may become completely impossible for some aphasic, paralyzed stroke patients, for victims of certain degenerative neurological diseases, and for those who are physiologically "locked in."

Of course, extreme old age is not always this grim, and it may well be objected to the argument just advanced that it trades on worst-case scenarios. Instead, it may be argued, old age may have many positive features, the precluding of which would be a substantial loss. Some of these features are reflected (albeit rather sentimentally) in what we might call the "Golden Pond" model of old age: an idyllic pastoral scene of two loving old people at home in a lakeside cottage, comfortable in means and content with their lives, whose sensory capacities remain vivid, whose minds are alert, and whose only physical problems are mild angina and a bit of a Parkinson's tremor. Similarly positive attitudes towards old age are portrayed in the poem quoted by Molly Gavin and Gayle Kataja earlier in this volume ([22], p.129, source not cited); I quote it not for its literary merits but to exhibit the heroic view of old age it recommends:

> Yes I am old – my strength declines
> And wrinkles tell the touch of time.
> Yet I might fancy these the signs
> Not of decay, but Manhood's prime
> For all within is young and glowing
> Spite of old age's outward showing.
>
> Yes, I am old, experience now
> That best of guides hath made me sage,
> And thus instructed I avow
> My firm conviction that old Age
> Of all our various terms of living
> Demands our warmest, best Thanksgiving.

Isaac Bell, 1857

In general, thus, the rational self-interest maximizer surveying the prospect of old age confronts two quite different, heavily stereotyped views: the grim, worst-case scenario depicted earlier, and the rosy Golden Pond model described now, reinforced by the kinds of attitudes recommended in Isaac Bell's poem. Both are extreme, and the actual conditions of life for most of the very elderly lie somewhere in between.

To view these two models simply as limiting conditions or best- and worst-case extremes, however, is to obscure the crucial feature here. As physical and/or mental deterioration progresses in old age, the role of the agent in determining which of these scenarios his life will more closely approximate also decreases, and it is this that is a centrally important fact: the circumstances one faces in extreme old age are very largely beyond one's own control. To be sure prudent financial planning and the development of healthy family relations earlier in life will play a major role, but nevertheless the factors that determine whether one's lot is the Golden Pond idyll or, on the other hand, grim institutionalization or dependent status in an adult child's home, is due in quite large degree to factors partially or entirely beyond one's own control: the onset of disabling physical conditions or chronic illnesses requiring continuous nursing care; the willingness or unwillingness of family members or others to undertake care; the attitudes of one's physician, especially at discharge from an acute care hospital, the policies of skilled care facilities, reimbursement policies for health care costs, and so on. However fiercely one may have protected one's autonomy at earlier stages of life, what happens to that same person in extreme old age is largely decided by second parties: family members, doctors, hospital and nursing home administrators, insurers, governmental policy makers, and others. In general, where you live is a function of which of your children will take you in, or what sort of institution you are sent to instead. What you can do, and where you can go, is a function of what those children will permit and help you to do, or whether it falls within the scope of an institution's official programs and rules. The costs to you and to your family may also be beyond your control, subject only to the limits of your or their absolute ability to pay. The general point is this: whether one is better or worse off in a lakeside cottage or in a nursing home, with one's family or at some remove, with constant medical treatment or without it, it is in extremely old age no longer a matter of one's own choice. One can neither predict one's actual future, at least not with certainty, nor control it in any very effective way at all.

Adding to the discomforting complexity of the prospect which the rational self-interest maximizer faces in surveying the possibilities of old age is the fact that the rosy and the grim versions are sequentially related. Sometimes a person may emerge from a period of extreme physical and/or mental disability into a comparatively rosy period, but for the vast majority of the elderly the sequential relation is the reverse: the person who achieves the Golden Pond idyll or some approximation of it moves perhaps gradually but nevertheless inexorably towards the grimmer picture as he ages – unless, of course, death intervenes first. Not all the elderly reach the grimmest extreme of this worst-case downhill course – and in emphasizing this fact the gray-power groups are entirely right – but what is rarely explicitly acknowledged is that the shift from the rosy to the grim model can be avoided only by death. The Golden Pond model, if it is achieved at all, cannot be sustained forever; sooner or later it is interrupted, either by deterioriation or death. In some cases, the onset of deterioration is initiated by an accident or traumatic event, such as a fall resulting in a broken hip; in others, the cumulative effects of slow deteriorative processes finally become pronounced. Of course, not all elderly people enjoying the Golden Pond idyll late in their lives deteriorate; some simply die without suffering decline. But it is important to understand here that what protects against decline is death itself.

These facts may make suicide attractive to the rational self-interest maximizer. The only sure route to self-determination in old age may seem to be to end one's life on one's own terms, at a time and place and in a manner of one's own choosing, before irreversible deterioration and late-life dependency set in. Viewed in this way, suicide is conceived of as a kind of self-defense,[2] protecting oneself (and one's family) from extremely negative outcomes. To be sure there is a price – the risk of foregoing some relatively small amount of remaining good life – but it is a price the self-interest maximizer may be willing to pay.

Such a choice may result from using a maximin strategy for decision under uncertainty: it ranks the foreseen alternatives for old age – from the grim worst-case scenario to the Golden Pond idyll – by their worst possible outcomes, and adopts as the alternative the worst outcome of which is superior to the worst outcomes of the others ([30], p. 152). It is better not to exist at all, the maximin strategist reasons, than risk the gruesome conditions of the grimmest picture, especially since once the onset of these conditions has begun, one will have lost virtually all opportunity to alter or avoid them. It is only by choosing suicide "before need," so to speak, that its self-protective value can be exercised. Of course, suicide would not be the choice of the maximax strategist, the person who ranks his choices by their best possible outcomes: he is willing to hold out for the Golden Pond idyll even though he realizes he may face a severe loss if his hopes are thwarted. Nevertheless, we may assume that rational self-interest maximizers surveying the prospects of extreme old age – that is, all of us who do not succumb to something else first – include both maximin and maximax strategists, and while the maximin strategy favoring suicide will not be congenial to all, it may be so for many. Even if the decision were to be made under risk, where the agent knows the probabilities of the various possible outcomes, the utilitarian calculus he might employ – multiplying the relative desirabilities or undesirabilities of various outcomes by the probabilities of their occurrence – will still yield the same prudential results if the worst outcomes are valued more negatively than death. In fact, it is not clear whether choices about old age are best construed as choices under uncertainty or risk: the agent may have some understanding of the probabilities of various outcomes in extreme old age (though many ordinary people are woefully misinformed), but very little understanding of how his own individual characteristics influence these probabilities or what the responses of others are likely to be; or it may be the other way around. Generally, we know something about the risks of old age, but not nearly enough for informed decision-making under risk. Consequently, I'll speak for the most part as if decisions about old age are decisions under uncertainty, though similar arguments could easily be advanced for decisions under risk. In either case, given that extreme old age has some very bad things to offer and no guarantees for avoiding them, protecting oneself by suicide – even at the cost of missing a period of additional life – may seem a reasonable prudential choice.

Perhaps still further enhancing its attractiveness as a self-protective strategy against the worst of old age, traditional moral objections to suicide seem to have little purchase when employed in such cases. For instance, that suicide may do severe emotional damage to family and friends, a fact painfully evident among younger suicides, is much less likely in extreme old age. The elderly person is unlikely to have dependents; his most intimate personal relationships, presumably with contemporaries, are likely to have been severed already by death, incapacitation, or institutionalization; and his death is not far off in any case. Similarly, the elderly person can no longer be blamed for failing to contribute his labor or talents to the welfare of the community, since – whether by reason of incapacitation, stigma, or forced retirement – he typically no longer works, and indeed may constitute an economic drain rather than resource. The elderly person can also no longer be blamed, as might be the case under natural-law conceptions of ethics, for failing to fulfill man's natural functions, since continuing deterioration will render these functions impossible in any case. Furthermore, suicide is not a violation of the law, contrary to widespread belief, and need not thus undermine the social fabric. Finally, that suicide might set a dangerous example to others is not a risk if the example is perceived as one of rationally self-interested, morally responsible suicide in old age, not applicable to younger persons who are depressed or emotionally distraught. Most of the traditional moral arguments against suicide emphasize its damaging effects for other persons and for society as a whole, but these effects are typically minimized, absent, or reversed in extreme old age. If suicide may be the choice of a rational self-interest maximizer and if it does no undue harms to others, there would seem to be little basis for objection to its practice in a routine, widespread way.

2. SUICIDE IN ECONOMIC CONTEXT

But individual choice always occurs within a larger economic context, and this context may change the moral coloration of the choice. The very possibility that suicide may prove attractive to morally responsible self-interest maximizers surveying the prospects for old age, and thus perhaps be practiced in widespread way as the preferred solution to the problem of old age, raises further ethical issues. These issues are concerned with the potential economic effects of the practice of suicide in old age, and raise basic problems in distributive justice.*

Let us put the root of the problem in the simplest, quickest way: *suicide is cheap*. Those reaching old age, and especially those entering extreme old age, are those for whom late-life dependency has or may become a reality, for whom medical care expenses are likely to escalate, and for whom needs for custodial and nursing care will increase. People over 65 use medical services at 3.5 times the rate of those below 65 ([13], p. 515). In 1981, the 11% of the population over 65 used 39.3% of short-stay hospital days, and the 4.4% over 75 used 20.7% ([2], p. 70). There are now about 6 million octogenarians, and the Federal Government

provides an estimated $51 billion in transfers and services to them
([33], p. 1). People 80 years of age or older consume, on average, 77%
more medical benefits than those between 65 and 79 ([33], p. 6).
Nursing home residents number about 1.5 million, of whom 90% are 65
or over, at an average cost of $20,000 per year ([32], p. 140). Although
only 4.7% of persons 65 or over are in nursing homes, rates rise with
age. About 1% of persons 65–74 are in nursing homes; of those 75–84,
7%, and of those 85 and over, about 20% are in nursing homes on any
given day ([32], p. 73). Even so, persons institutionalized represent a
comparatively small fraction of the elderly suffering chronic illnesses
and disabilities, and it is estimated that for every nursing home resident,
there are two other people with equivalent disabilities in the community
([31], p. 831). One estimate suggests that 70% of the elderly who need it
rely on relatives for care ([32], p. 141). Even if a person maintains
functional independence into old age, the risk of becoming frail for a
prolonged period is still high: for independent persons between 65 and
69, one study found, total life expectancy was 16.5 years, but "active life
expectancy," or the portion of the remaining years that were characterized by independence, was only 10.0 years, and the remaining 6.5 years
were characterized by major functional impairment. Furthermore, this
risk increases with age: persons who were independent at 85 were likely
to spend 60% of their remaining 7.3 years requiring assistance ([31,
quoting Katz et al.], p. 828). Expenditures are particularly large for
those who are about to die: for instance, for Medicare enrollees in 1976,
the average reimbursement for those in their last year of life was 6.6
times as large as for those who survived at least 2 years, and although
those who died comprised only 5.2% of Medicare enrollees, they
accounted for 28.2% of program expenditures ([28], pp. 6–7). Further,
a report from the National Center for Health Statistics shows that 62%
of persons 65 or older die in a hospital or medical center (quoted in [21],
p. 162). The resources in time, effort, and money devoted to the care of
the elderly are enormous, and suicide undertaken at the onset of
dependency and physical or mental deterioration requiring extensive
medical or custodial care would have major impact on the distribution
of these resources. The problem looms in this stark way: suicide is the
cheapest solution to the costs of dependency for those who reach
extreme old age.

   This is not a comforting fact. But it is not possible to understand the
central issues about the special moral problems of suicide in old age

unless this fact is recognized and understood. To see the philosophical problem at all, we must resist our very natural, well-intentioned, humane inclination to deny that the elderly may constitute an economic and social burden: instead, we must force ourselves to see that the medical and custodial care of the elderly involves an extremely sizeable expense, and that the frequent choice by the elderly of suicide before the onset of such problems would markedly reduce this load. This is true both for expenses to the immediate family, including both the financial costs of treatment and equipment and non-dollar costs such as time spent in care, and for expenses to society as a whole. Indeed, in a society acutely aware of the escalation of medical and custodial costs, and in which the proportion of elderly persons in the population as a whole is increasing very rapidly, the widespread acceptance of suicide as an alternative to deterioration and dependency in extreme old age would prove perhaps the most effective cost-containment measure possible. It might not, of course, be favored by the medical establishment itself, since a sizeable portion of physician and especially hospital revenues comes from these sources, nor by the nursing home or life-insurance industries, but in a society concerned to control the overall proportion of expenses devoted to medical and related care, such practices might prove attractive indeed.

These claims suggest that there are powerful financial incentives favoring suicide in old age, and that if perceived as in concert with the self-interests of the elderly and as morally acceptable, a cost-conscious society eager to spare itself burdens might actively favor such choices. This is not to say that such choices would be enforced, nor that they would be required of those who did not make them freely; but it is to speculate – not prove – that societal expectations encouraging suicide in old age might well develop. Such expectations need not take the form of legal or social requirements, but might instead be the kind of easy, natural expectations which develop within a social environment when it recognizes certain sorts of behavior as prudent and commendable. I have in mind the sort of general expectation which encourages, say, marriage in early adulthood, or (depending on one's social class, ethnic background, etc.) going to college, defending one's territory, or displaying one's wealth in reasonably specifiable ways. We may speculate, thus, that a consequence of permitting suicide at all for those who wish to protect themselves against the worst that old age has to offer would be the development of favorable attitudes towards the possibility of elderly persons in general doing so. With increasing acceptance, the prescriptive force of such attitudes might well increase in strength. Eventually, perhaps, what had been an option for some would become an expectation for all, and it would be widely held that this is what the elderly *ought* to do.

This suggestion may rightly be recognized as a form of wedge or slippery-slope argument, predicting that if suicide choices by maximin strategists among the elderly are countenanced to begin with, expectations of suicide will eventually be directed toward everyone who survives into old age: it will come to be accepted and encouraged that choosing one's time to die before the onset of profound dependency in extreme old age is one's social and moral obligation. Such a choice might of course be coerced by direct threats or deprivations – suppose, for instance, that pension funds and social security simply ceased to provide benefits after age 85 or 90 – but it is also possible to imagine social policies which do not involve coercion of any flagrant kind, since such choices can be encouraged by manipulation of the most subtle sort. After all, it is society which ultimately determines the criteria of rationality by defining the parameters that are to count as important: completing projects, having human communication, and so on, and thus can dictate the point at which the rational self-interest maximizer assesses himself as better off dead. The criteria of rationality can be altered by manipulating attitudes and values ([6], pp. 160–175), and we may expect this to occur where economic incentives strongly favor an earlier, less expensive death. We can foresee the newly advertised and accepted view of suicide: "It's what you do when you get too old." This is particularly plausible if society begins to develop suicide-supportive mechanisms, such as legalization of physician assistance in suicide, living wills directing not merely cessation of life-prolonging treatment after the patient becomes incompetent but direct intervention to produce death at a specified point, and changes in social and religious attitudes about suicide. It is even more plausible if the scope of suicide choices is understood to include voluntary acquiescence in no-treatment and no-code decisions made by physicians and others: though the patient might ostensibly be given an opportunity to object to such decisions, it would be expected of him that he not do so. We have no firm evidence that any of these outcomes would occur, of course, but because of their gravity, very good reason to take such possibilities into account.

Although most wedge arguments are best addressed either by challenging the causal assumptions that they employ or by showing that it would be possible to halt the slide down the slippery slope before the unacceptable outcome occurs, in this case we may take the wedge argument to pose a rather different sort of challenge.[3] Suppose the acceptance of suicide in old age were to lead to societally-fostered expectations that the elderly choose their times to die, and not prolong their lives until the medically inevitable end. Such expectations would of course be imposed not just on maximin strategists, who might choose suicide as a form of self-protection in any case, but upon all the members of the culture as a whole, including maximax strategists and others who might in the absence of such expectations choose to stay alive. Would such expectations be morally repugnant, as we tend to assume? The question for us is not the empirical one of whether such expectations would develop, but whether, if they did so, there would be any basis for moral objection to them. I view this as a question heavily tied to basic distributive issues, made compelling by the economic context within which it arises.

3. AGE-RATIONING AND SUICIDE

In the fifth century B.C., Euripides addressed "those who patiently endure long illnesses" as follows:

> I hate the men who would prolong their lives
> By foods and drinks and charms of magic art
> Perverting nature's course to keep off death
> They ought, when they no longer serve the land
> To quit this life, and clear the way for youth.

(Quoted in [6], p. 99.)

More recently, controversy has surrounded former Colorado Governor Richard Lamm's alleged remark that the ill elderly have a "duty to die" (see his own account, [27], pp. 20–23). Such expressions would seem to assert that societal expectations urging suicide are not morally repugnant, and indeed are something for which there is clear moral support. Yet is this correct? These views raise crucial questions of distributive justice about whether the elderly are entitled to equal access to medical, custodial, and other care, or whether advanced age weakens any claim they may have had. After all, if the elderly have only a diminished claim to health care or perhaps no claim at all, the social encouragement of suicide in old age could be understood as a mechanism for effecting age-rationing requirements which, in a scarcity situation, justice may appear to demand.

Assuming the underlying formal principle of justice to require that like cases and groups be treated alike, we may consider whether material principles of justice would differentiate the elderly from other claimants for health care. But it is not clear that this strategy will succeed. One might argue, for instance, that entitlement to health care is a function of the contributions society may expect as a return on its investment; on this claim, since the elderly can no longer make contributions, such support would not be justified. But the elderly have already made their contributions, and these are in fact more secure than those of younger claimants whose contributions remain so far only potential; on this counterargument, society owes the elderly care in return for the contributions they have already made. Similarly, one might argue that the elderly have greater claims to care in virtue of their greater vulnerability, in virtue of the respect owed elders, or in virtue of the intrinsic value of old age. This sort of discussion, characteristic of most analyses of distributive justice, involves identifying the possible desert bases in claims to health care, and then considering whether the elderly can satisfy these conditions as well as other age groups. If they do (which I think likely), policies which restrict the access of the elderly to certain kinds of care, now evident in England ([1] *passim*) and perhaps as well here, must be seen as the products of simple age bias.

But an influential observation has been made by Norman Daniels ([13] *passim*). This conception, Daniels suggests, should be modified by recognizing that the elderly are not another age group, distinct from, say, infants, adolescents, and the middle-aged, but that they are to be viewed as the same persons at a later stage of their lives. Our mistake is in considering distributive problems as problems in allocating resources among competing groups and among competing individuals, when we more correctly would consider the problem as one of allocating resources throughout the duration of lives. Daniels employs Rawlsian strategies to make this point: he invites us to consider what distributive policies we as prudential savers would adopt if, unable to know our own medical condition, genetic makeup, physical susceptibilities, environmental situation, health maintenance habits, or age, we must decide in advance on a spending plan budgeting a fixed amount of medical care across our whole lives. He conjectures, and I think plausibly so, that rational self-interest maximizers behind the veil of ignorance in this original position would choose, where scarcity obtains, to allocate a greater amount of resources to care and treatment required for conditions that occur earlier in life, from infancy through middle age, but not to underwrite treatment which would prolong life beyond its normal span. By freeing resources now devoted to prolonging the lives of the elderly for use instead in the treatment of diseases which cause death or opportunity-restricting disability earlier in life, such a policy would maximize one's chances of getting a reasonable amount of life within the normal species-typical, age-relative opportunity range. If it is a policy on which prudential savers would agree, Daniels holds, it will show that – at least under scarcity conditions against a background of just institutions – age-rationing is morally warranted for making allocations of health care.

But let us push an account justifying age-rationing to the extreme. An account of the sort Daniels gives, for instance, would invite prudential savers to calculate the chances of late-life deterioration and dependency, which would require extensive medical and custodial care, against the chances of birth abnormalities, childhood diseases and accidents, and various mid-life illnesses, all of which also require care. These original-position savers do not of course know their own individual ages, health statuses, or habits, but, at least under an appropriately thin veil of ignorance, they will know the statistical probabilities of these occurrences in general. (Daniels considers both thick and thin veils.) Furthermore, not only will they recognize that preserving life in the earlier years is absolutely prerequisite to having life in the later ones, and that for the most part preserving function in the earlier years is also prerequisite to maintaining function in the later ones, but they will also know that because older persons typically have more complex medical problems, compounded by decline in the function of many organs and by reduced capacities for healing and homeostasis, tradeoffs between earlier and later years cannot be made on a one-to-one basis: by and large, a unit of medical care consumed late in life will have much less effect in preserving life and maintaining normal species-typical function than a unit of medical care consumed at a younger age. They will know, in other words, that although the chances of ill health are greater in later years, a dollar's worth of medical care cannot buy nearly as much good health and normal functioning in these later years as it could have made possible in earlier ones. For this reason, the prudential saver may well be inclined to allocate his lifetime share of health care in such a way that it strongly favors the earlier years, when, so to speak, he can get his money's worth; if reserved for the end of life, his share will provide very rapidly decreasing benefits to him. If the veil of ignorance is very thin, for example, the prudential saver will recognize that the $9437 of total hospital costs a non-survivor (in 1979) would incur during his last 13 days in the hospital and 5 days in an intensive care unit ([16], p. 669) could provide extraordinary benefits to him if used instead in early years to provide pre- or post-natal screening and nutrition, life-saving or function-improving therapy or surgery for birth abnormalities, prompt repair of trauma for adolescent accidents, or careful monitoring of mid-life hypertension or diabetes. On the other hand, if it is spent in old age, after the onset of multiple chronic illnesses and progressive deterioration, it does little more than prolong dying, and can hardly be said to maximize the normal, species-typical age-relative opportunity range. What the prudential saver will be willing to forgo, if forced by scarcity to sacrifice something, is the fruitless, expensive decline, characterized by extreme dependency and medical deterioration, at the very end of life: he will trade off his oldest years for better chances of survival and function in his younger ones, recognizing that this choice in effect disenfranchises him from any claims to more than minimal, superficial health care at the end of life.

Of course, what the genuinely prudential saver will favor is not an age-rationed distributive policy, but a time-to-death-rationed policy. Variations in costs and efficacy of treatment are not so much a function of time since birth, but time to death ([21], pp. 151–2); it is the last month, six months, or year of life that proves expensive, not all the years above a fixed age. Some – indeed many – octogenarians are vigorously healthy; so are some people in their nineties and beyond. On the other hand, dying can be expensive, and medical efforts essentially futile, even for those whose ages are not advanced. Thus, the prudential saver would not favor an age-rationed policy, but rather one which, depending on the degree of scarcity, disenfranchised him only during the last month, half-year, or year of life. Of course, the precise ante-mortem period can be identified with certainty only retrospectively, and it is this that contributes to the problem; prognoses and predictions of remaining survival time are often unreliable. No doubt it is possible for the experienced physician to recognize the onset of what is likely to be a downhill course ending in death with at least a fair degree of accuracy; even if such predictions are sometimes inaccurate, however, the prudential saver will still prefer reliance on them, in order to maximize his opportunities for continuing life and normal functioning, to rigid age ceilings. Furthermore, since some declines are comparatively rapid and some prolonged, the prudential saver will seek to maximize his overall opportunities not by agreeing to a policy in which a fixed amount of time at the end of life is held ineligible for care, that is, an ante-mortem period of uniform and predetermined duration, but by supporting a policy in which disenfranchisement begins only at the onset of profound dependency and irremediable chronic disease. This may mean forgoing a couple of weeks in an ICU or a couple of years of severe incapacitation in a nursing home; either way, the point of disenfranchisement will occur neither solely as a function of time from birth or of time to death, but as a function of dependency and irreversible deterioration. Nevertheless, neither time from birth nor time to death can be ignored in assigning the point at which disenfranchisement would begin: in general, such policies would apply only to those of advanced ages, and not to much younger persons who are seriously ill, nor would they apply to persons who might expect many years of chronic illness where life nevertheless continues to be a boon. Prudential savers, in short, will tag the point of disentitlement neither to a fixed age ceiling nor to a fixed point of time-to-death, but rather to the onset of severely limiting acute or chronic medical conditions which will very shortly result in profound dependence, severely limited opportunity ranges, and sharply increased healthcare needs. It is the final illness, whether swift or extended, they will elect to avoid, neither old age *per se* nor a fixed period of years.

But if skewed allocations of this sort form a general policy to which prudential savers would agree, thus providing warrant for rationing policies which, while taking into account both age and life expectancy, disenfranchise elderly persons at the onset of severe late-life dependency, we may also press these prudential savers to stipulate how they would wish to be treated when they do reach old age and fall ill – given that they subscribe to policies under which they have allocated their healthcare resources to other, earlier ages. We may take whatever policy these same prudential savers would again agree on as warrant for policies to be developed in practice, and as a basis for determining how in the real world we ought to respond to the elderly ill. Rationing policies, whether based on age, time-to-death, or other factors, are always morally incomplete without attention to these disturbing questions at the end: what ought to be done, if anything, for those who become disenfranchised in this way?

If prudential savers adopt a policy disenfranchising themselves from health care in the final stages of life, two – and only two – basic options are open to them. They can on the one hand simply agree to go without treatment – that is, accept a policy holding that they are appropriately denied it – or they can consent to end their lives at or before the time at which treatment is required, that is, they can adopt policies expecting them to commit suicide. There are, of course, many varieties of these two basic options: they can be content with very limited home hospice care (not to be confused with the much more expensive inpatient hospice care) and minimal palliative treatment; these policies would still fall under the general rubric of foregoing (maximal) care. Or they could institute physician-assisted voluntary euthanasia; this would count as a variety of suicide. Both are cheap, and both would satisfy the allocative priorities agreed to by prudential savers under which the bulk of an individual's lifetime health care share was allocated elsewhere.

Yet these two basic responses to the disenfranchisement problem may have quite different effects both for the persons involved and for society as a whole, and the social policies emerging from these choices may thus be quite different in character. Thus, we may pose the problem from a practical perspective: if the elderly are disentitled to medical and custodial care, ought our society to deny them treatment – as is now beginning to be the case – or encourage and assist them, as many primitive societies have done, in ending their lives?

4. RESPONSES TO DISENFRANCHISEMENT

Describing a practice attributed to the Aleuts, Daniels says:

The elderly, or the enfeebled elderly, are sent off to die, sparing the rest of the community from the burden of sustaining them. From descriptions of the practice, the elderly quite willingly accept this fate, and it is fair that they should ([13], p. 513.)

In various migratory American Indian groups, old, ill persons who were too frail to keep up were abandoned by the side of the trail; the early Japanese, at least according to legend, took their elderly to the mountaintop to die; during the siege of the Greek island of Ceos, a law was passed requiring persons reaching the age of 65 to commit suicide, though this law was rescinded when the siege was over. Such practices may be understood as responses to culturally accepted views about comparative entitlements to care, according to which the claim of the elderly ill weakens under scarcity conditions, as for instance in the marginal economic conditions of these primitive or beleaguered societies.

To imagine contemporary situations incorporating the same response, that is, in which suicide is socially encouraged as a solution to the scarcity problem, is not at all difficult; indeed, the wedge argument predicts the development of just such expectations. We have already seen that, for the very elderly, suicide can be a rational choice, independently of any considerations of justice, because it allows one to retain control in preventing worst-case outcomes. We have conjectured that social support for suicide could be generated in a variety of ways: by praising persons who make this choice, by providing medical assistance, by developing social and religious "farewell" ceremonies, and so on. Furthermore, under such expectations, many of the traditional objections to suicide would evaporate, especially those pointing to depression and mental illness as factors which impair the suicide's judgment. In social circumstances in which suicide is expected, miscalculation and manipulation will not in general be a problem, since the expectation carries with it a fairly clear indication of the time at which it will be appropriate to act on it – just prior to the onset of irreversible late-life dependency – and provides for expert guidance and protection from physicians and others in determining when this moment is reached. Society would help, so to speak, in choosing the time to die.

It is true that such expectations might be coercive, but moral critique cannot assume that this would be the case. We are familiar with many other social supports and expectations which also control behavior; however, not everyone conforms to such expectations, and where the possibility of non-conformity exists we continue to perceive behavior in accord with these expectations as voluntary. In this culture, as we've said, young adults are expected to marry, and most of them do; nevertheless, marriage is perceived as a self-determined, voluntary choice in part because those who choose not to marry are not coerced into doing so. Cultural expectations encouraging suicide might similarly respect individual choice and refrain from coercing non-compliant persons, rejecting the compulsory "death parlors" envisioned by the media, yet still be strong enough to be effective in the large majority of cases. Strong enough expectations sufficiently well internalized by the population on which they operate might thus both achieve the necessary effect and still permit satisfaction of individual preferences; indeed, it might well be possible to provide maximal medical care for persons who chose to resist these expectations and continue their lives, since if most people were compliant in choosing an earlier, cheaper time to die, the savings of resources their choices would yield would be quite adequate to provide the redistributive effect that justice demands. (Of course, one might object on moral grounds to free riding, but infrequent choices of this sort would not constitute a practical problem.) All that the expectation need convey, to be effective, is the notion that there is such a thing as being too old, especially where debility means dependency, and that it is both prudent and morally responsible to prevent this from occurring; indeed, this notion might be as heavily promoted and widely accepted as the notion that life is of absolute value now is. To convey such a notion, and to provide assistance in acting on it, is one way in which society could respond to a rationally supported distributive policy according to which the elderly are no longer entitled to substantial health care.

No doubt society could respond in this way. No doubt, too, such a response would be adequate to satisfy the demands of justice. But this is not yet to defend such a response; indeed, to allow or encourage expectations to develop that old, ill people kill themselves in order to resolve distributive problems may seem repugnant indeed. Specifically, this is not yet to establish that this response is morally preferable to the only major alternative that disenfranchisement offers: this is the policy of denying the elderly treatment, at least where the costs of treatment are substantial. Under a denial-of-treatment policy, "cheap treatment" such as common antibiotics could be made available for elderly patients; but expensive diagnostic procedures and therapies like CAT scans, dialysis, organ transplants, hip replacements, hydrotherapy, respiratory support, total parenteral nutrition, individualized physical therapy, vascular grafting, major surgery, and hi-tech procedures generally would be ruled out. Hospitalization, and the nearly equally expensive inpatient hospice care, might not be permitted, except perhaps briefly; sustained nursing home care would probably also be excluded. The elderly person over an appropriate age ceiling or who exceeds a pre-determined level of deterioration who now shows symptoms of a more than quite transitory, easily cured illness would simply be counted ineligible for treatment. "I'm sorry, Mr. Smith," we can expect the physician to say, "there is nothing we can do."[4] This may initially seem a much less disturbing response to the distributive problem than encouraging suicide, even though in both cases the result for the patient is an earlier death than maximal treatment might have allowed.

Unexamined reactions to these policies, however, will hardly suffice for ethical scrutiny; rather, we must ask whether there are clear moral considerations favoring this policy of denying treatment over the alternative policy of encouraging suicide – given that one or the other is necessary to satisfy the rationing requirements of justice – or whether it is the other way around. There are, I think, three major considerations to be made. One favors policies of denying treatment, but the other two favor encouraging suicide.

First, considerations of the value of life seem to favor denying treatment, since this will allow the patient to attain the maximum amount of life possible without further medical care. Under the denial-of-treatment model, death occurs as the person succumbs to his illness in the absence of medical care; under the expected-suicide model, death would take place somewhat prior to this, at the onset of the illness. But since in any case the effect of rationing is to shorten life in old age, albeit to maximize chances of life and function at earlier points in the lifespan, considerations of the value of life *per se* can hardly be primary here.

Secondly, considerations of mercy seem in contrast to favor the expected-suicide model. This is because the extra period of life the denial-of-treatment model offers the patient may not be a benefit, since it is time during which he suffers his illness but goes untreated. Of course, minimal home hospice care and inexpensive pain relief could be made available, but more expensive forms of treatment would not. Yet many of the kinds of treatment which the elderly patient would thus be denied do a great deal to relieve symptoms, promote function, and restore comfort. A total hip replacement, for instance, may serve to extend life expectancy, but it does so in part by improving mobility, and the difference in the character of life for an elderly person no longer able to walk and one who remains ambulatory can be quite great. The person who needs, but does not get, a pacemaker or a coronary bypass may lead a quite restricted life, seriously limited in his activities. Life with renal failure, or cardiac arrhythmias, or pulmonary insufficiency can be uncomfortable, frightening, or painful. Allowing the patient to succumb to his diseases because treatment is no longer provided may be to abandon him to considerable physical and psychological misery, and may make still more awful that catalogue of worst-case scenarios with which we began. Even symptom control, if not merely obtunding and completely obliterative of consciousness, can be quite expensive, since effective control may require constant titration and monitoring; if so, it too would be ruled out. Old old age is not always pleasant, and to deny a person treatment for the conditions he develops may make his remaining time less pleasant still. Worse still, common antibiotics and the few other kinds of cheap treatment that are available may simply serve to prolong this period, not to reduce its discomforts. The denial-of-treatment model does not simply bring about sooner deaths that maximal care would postpone; denial of treatment also means denial of expensive palliative measures (e.g., reductive surgery) which maximal care would permit. It is of course true that under the denial-of-treatment model the patient might suffer his illness without treatment and nevertheless survive, but with increasing age the likelihood of this outcome decreases. In old age, one must remember, it is not a choice of life vs. death, but death now vs. death a little bit later. Death and the decline which precedes it may be easier to endure, whether sooner or later, if lavish care is provided rather than if all but minimally expensive care is denied. To simply deny treatment may be to increase the risks of worst-case outcomes, the very thing the rational self-interest maximizer may wish to avoid (see [7], *passim*).

Thus, while considerations of the value of life favor simply denying the elderly ill care, considerations of mercy would seem to recommend encouraging voluntary termination of life before the onset of serious irreversible illness. Maximin strategists, anxious to avoid the worst-case outcomes, will welcome this latter recommendation in any case, though maximax strategists – long-shot takers that they are – will not. Now it may seem that the choice of policies reduces to a choice between the preferences of maximin and maximax strategists, and the differences between these decision styles, in turn, may seem simply to be a matter of individual discomfort-budgets: how much discomfort, how much misery, is one willing to tolerate in exchange for how great a chance of more life? Since the impact of denial-of-treatment and expected-suicide policies appears to have very nearly equivalent consequences for distributive justice, and since neither considerations of mercy nor of the value of life seem decisive in controlling a patient's choices, it may seem to make little moral difference whether denial-of-treatment or expected-suicide policies are adopted if the elderly ill are to be disenfranchised. Of course, the former does much less violence to our conventional attitudes, and thus might on that weak basis be preferred; but I do not think the matter is quite so simple as that.

This is because there is an additional problem with the denial-of-treatment model which does not arise with the expected-suicide model, a problem which has sizeable consequences for the issues in justice. This third consideration I like to call the "erosion" problem, short for "the erosion of justice by mercy"; it has to do with the stability of expectations and the moral position the denial-of-treatment model places health-care providers, family members, policy makers, and other on-lookers in. Under the denial-of-treatment model, an elderly person who falls ill is to be denied all but minimal care. But as we've seen, the denial of care can have quite cruel consequences. The caregiver, observing this, can hardly be expected to enforce this policy; after all, as a caregiver, he operates under the canon of mercy, not as an officer of justice. To require the caregiver to observe the denial-of-treatment model is to require him to abandon his patient in a way that runs counter to the most basic principles of his professional commitment. Further-more, although some patients may wish to forego mercy in order to maximize their chances of living – this is the choice of the maximax strategist – under the distributive policy considered here the physician cannot offer them care to enhance these chances. Thus, under the denial-of-treatment model, the physician is placed in an intolerable position with respect to both patients who are maximin and maximax decision-makers: on the one hand, he finds it inhumane to deny treat-ment to his patient and then insist that the patient suffer these reduced conditions, especially when the reduction in comfort and function is severe; on the other hand, he finds it insincere to allow his patient to take the longshot chance if he is prevented from doing anything to help. Either way, the practical impact on the physician or other caregiver is clear: his whole effort will be to respond to the demands of mercy, to look for loopholes in the rationing policy, for pretexts under which he can provide either genuinely effective palliation or the aggressive care his patient wants. In practice, thus, the denial-of-treatment policy would prove unstable. But every erosion of the policy by providing extensive palliative or aggressive care violates the requirements of justice and the spe-cific rationing policy they establish. Furthermore, the erosion problem is compounded by the fact that medical gains, whether in preservation of function or prolongation of life, are very much more expensive to achieve in old age, since, as we've pointed out, the body's natural healing and homeostatic capacities are much reduced. The erosions of justice by mercy that might be cheap in youth are costly in old age, and grow increasingly so. Under the denial-of-treatment model, the requirements of mercy and the demands of justice are at constant odds.

It might seem to be the case in the expected-suicide model too that the requirements of justice would conflict with those of mercy. But, if historical practices from other societies are any guide, we must remember that under the expected-suicide model, suicide may be voluntarily, even willingly embraced in response to supportive social expectations. The rationality of choosing one's time to die is pointed out; the moral blamelessness of this choice is pointed out too, and ceremonial recognition is accorded (perhaps under the authority of the churches) in which the person is praised for the responsible decision he is about to make. It need not involve violence or bloodshed, but a quiet, controlled, socially respected end. Suicide is simply seen as the natural, preferred, customary end of life, much to be preferred to a slow, deteriorating decline and to waiting, as Mary Rose Barrington puts it, to be "arbitrarily extinguished" ([4], p. 99). Thus, the caregiver who supports this choice and assists his patient in carrying it out is not violating the requirements of mercy, at least in the absence of any principle that death itself is always a harm. The support in death the caregiver brings is perceived as a benefit by the person to whom it comes, preferable on all counts to that abandonment he would otherwise face. Furthermore, such an expectation, operating on a society-wide scale, might also succeed in recognizing patient autonomy, particularly for those maximax strategists who wish to take longshot aggressive-treatment chances, since, as we said earlier, if a large enough majority of elderly persons respond to the prevailing expectation, savings adequate to effect the redistributive demand will result, and it would be possible to allow maximal, expensive care to those few who want it nevertheless. Since the situation is a longshot one, since the culture would no longer encourage taking such chances, and since in any case the longshot strategy increases the risk of worst-case outcomes, it may be supposed that relatively few self-interest maximizers will want maximal care. If so, their choices can be protected after all.

In view of these considerations, it becomes clear that the choice between the denial-of-treatment and expected-suicide models is not merely a matter of individual discomfort-budgets, to be enshrined in policy by following the majority preferences or by seeking to do least violence to unexamined moral feelings. Rather, the moral consequences of these two models must be considered, as well as the likelihood that either could be maintained. The denial-of-treatment model involves profound conflicts between mercy and justice – as we now see emerging both in England and in the U.S. – and because of these conflicts is both morally suspect and is likely to be unstable in practice. The expected-suicide model, at least once it is in place, would involve no such conflicts and presumably would be stable over a long period of time.[5]

Of course, in either distributive policy, whether it favored denying treatment or encouraging suicide, there would be enormous room for abuse. But our first project here is to consider which sort of policy is to be defended on its merits alone within the context of an ideal theory, and only later consider whether the background institutions are sufficiently just to permit its introduction and whether adequate safeguards could be erected to guard against abuse. No doubt a policy of encouraging suicide would invite abuse, especially against a background of less than just institutions; but consideration of these evils is not the project here. It is our project, rather, to establish whether the stronger moral case can be made for such a policy, or for its only possible alternative, in the first place.

5. OTHER OPTIONS

However, even if it appears on both moral and practical grounds that the expected-suicide model is to be preferred to the denial-of-treatment model as a solution to the problem of disentitlement of the elderly, this is still not to establish that either is better than a policy in which full medical treatment is made available to the elderly who wish it. There are several ways to address this issue. First, it may be the case that on more careful scrutiny, principles of justice which seem to support age or time-to-death rationing policies cannot be successfully defended; it may be possible to show that both the denial-of-treatment and the expected-suicide models are age- or illness-biased in unwarranted ways, even within a general Rawlsian scheme. Secondly, it may as a practical matter be possible to remedy the abysmal conditions the elderly risk, especially in the less pleasant nursing homes and more hostile domestic situations,

186 MARGARET P. BATTIN

so that worst-outcome scenarios seem less grim after all. And finally, it may be that the scarcity assumption which gives rise to problems of justice in the first place is not accurate. After all, in contemporary western society, unlike the Aleuts and the Greeks under siege on the isle of Ceos, scarcities in medical resources are artifacts of distributive choices among various kinds of social goods, including education, defense, welfare, art, and so on. What is indicated, of course, is some reinspection of the way in which we have come to assume that scarcity characterizes our own society as well. If both abandoning the elderly and expecting them to put an end to their own lives seem to be morally undesirable responses to disenfranchising them, we must remember that the need for either one can be obviated by altering our other priorities in a way that makes adequate health care available to all.

ACKNOWLEDGEMENTS

I'd like to thank my colleagues Leslie Francis, Tim Smeeding, Tom Reed, and Virgil Aldrich for comments on various drafts of this paper, as well as Ken Howe and other members of the journals club in the Medical Humanities Program at Michigan State University for providing me with a tape of their discussions of a draft.

University of Utah
Salt Lake City, Utah

NOTES

* Part of the argument of this paper are reiterated in my later paper, 'Age Rationing and the Just Distribution of Health Care: Is there a duty to Die?', *Ethics* **97** (Jan. 1987), 317–340.
[1] I take refusal of life-prolonging treatment to be properly counted as suicide when it is initiated by the patient with the intention of bringing about his own death, not merely avoiding treatment he does not want. I am persuaded, however, that a rigorous set of necessary and sufficient conditions cannot be produced to define "suicide" ([34], pp. 39–47), and I recognize that this definition will not coincide with the linguistic intuitions of many. Because I shall be talking primarily about suicide which involves direct termination of life, nothing much hangs on whether death-intending refusal of treatment is also counted as suicide or not.
[2] The notion of suicide as self-defense was first suggested to me by Tom Reed.
[3] I have argued elsewhere ([8], pp. 28ff) that the wedge argument is not successful if it fails to take account of the interests and rights of those whom the initial practice would favor – in this case, those maximin strategists who prefer suicide to risking the worst of old age.

This is just to say that in analyzing a wedge argument, it must be remembered that the initial practice and the practice which is predicted to develop have distinct outcomes, and that the negative consequences of not having the initial practice occur must be considered as well as the negative consequences of having the subsequently developed practice occur. I think, on balance, that this requirement tilts the argument in favor of permitting suicide, at least in specific terminal-illness and extreme old age circumstances envisioned here, but this is not the kind of objection to the wedge argument I want to pursue in this case.

[4] I'd like to quote verbatim, because it is so illuminating, a brief passage from *The Painful Prescription* ([1], pp. 36–37), describing the clinical mechanisms of age-rationing: "British physicians are candid about the way they discourage patients from insisting on dialysis. Asked how he would explain to her family the prospects of a sixty-five-year-old woman with kidney failure, one general practitioner first told us that he did not think it was up to him to decide whether she should be dialysed, that he would leave the decision to the consultant. But then he added, 'Obviously the patient is sixty-five and therefore does not come within the regional dialysis program.' When pressed on whether he might save everyone time and anguish by discouraging referral, he described how he would talk to the family. 'I would say that mother's or aunt's kidneys have failed or are failing and there is very little that anybody can do about it because of her age and general physical state, and that it would be my suggestion or my advice that we spare her any further investigation, any further painful procedure and we would just make her as comfortable as we can for what remains of her life.' Remarkably, few of the criteria for rejection are explicitly stated. Age, for example, is not officially identified as an obstacle to treatment."

[5] We may then ask why expected-suicide policies in historical and primitive cultures have not survived. In many cases, the answer points to the intrusion of Christianity, with its severe disapproval of suicide, into suicide-accepting cultures. Christianity brought to an end suicide-approving Stoic hegemony in Greece and Rome; the intrusions of Christian missionaries into non-European cultures, like the American Indians and the Aleuts, disrupted traditional beliefs. Japan was forced to make suicide illegal after its defeat by the Christian U.S. in World War II. Then too, in some cultures what had been a severe scarcity situation eased; this was true of the Greek island of Ceos. Finally, of course, such practices may not have survived because these cultures did not develop and defend a considered theory of justice holding that the elderly have only diminished claims to health care and other social goods, but merely practised discrimination against their weaker members.

BIBLIOGRAPHY

1. Aaron, H. J., and Schwartz, W. B.: 1984, *The Painful Prescription: Rationing Hospital Care*, The Brookings Institution, Washington, D.C.
2. *Aging America, Trends and Projections*, Second Printing 1984. Report prepared by the U.S. Senate Special Committee on Aging in conjunction with the American Association of Retired Persons. No publisher given.
3. Avorn, J.: 1984, 'Benefit and Cost Analysis in Geriatric Care. Turning Age Discrimination into Health Policy', *New England Journal of Medicine* **310**, 1294–1301.
4. Barrington, M. R.: 1969, 'Apologia for Suicide', in A. B. Downing (ed.), *Euthanasia and the Right to Death*, Peter Owen, London.

5. Battin, M. P., and Mayo, D. J. (eds.): 1980, *Suicide: The Philosophical Issues*, St. Martin's Press, New York.
6. Battin, M. P.: 1982, *Philosophical Issues in Suicide*, Prentice-Hall, Englewood Cliffs, N. J.
7. Battin, M. P.: 1983, 'The Least Worst Death: Selective Refusal of Treatment', *The Hastings Center Report* **13**, 13–16.
8. Battin, M. P.: 1987 'Euthanasia and the Demands of Mercy, Autonomy, and Justice', in T. Regan and D. VanDeVeer (eds.), *Health Care Ethics*, Temple University Press, Philadelphia, Penna.
9. Bayer, R. *et al.*: 1983, 'The Care of the Terminally Ill: Morality and Economics', *New England Journal of Medicine* **309**, 1490–1494.
10. Bedell, S. E., and Delbanco, T. L.: 1984, 'Choices about Cardiopulmonary Resuscitation in the Hospital: When Do Physicians Talk with Patients?' *New England Journal of Medicine* **310**, 1089–93.
11. Buchanan, A.: 1983, 'The Right to a Decent Minimum of Health Care', in President's Commission for the Study of Ethical Problems in Medicine and Biomedical and Behavioral Research, *Securing Access to Health Care*, Vol. 2, Government Printing Office, Washington, D.C., pp. 207–238.
12. Childress, J. F.: 1979, 'Priorities in the Allocation of Health Care Resources', *Soundings* **62**, 256–274.
13. Daniels, N.: 1983, 'Justice between Age Groups: Am I my Parents' Keeper?' *Milbank Memorial Fund Quarterly* **61**, 489–522.
14. Daniels, N.: 1985, 'Equal Opportunity and Health Care Rights for the Elderly', paper presented to Conference on Rights to Health Care, University of Missouri School of Medicine.
15. Daniels, N.: 1985, 'Is Age-Rationing Just?', paper presented to Pacific Division, American Philosophical Association.
16. Detsky, A. S. *et al.*: 1981, 'Prognosis, Survival, and the Expenditure of Hospital Resources for Patients in an Intensive-Care Unit', *New England Journal of Medicine* **305**, 667–72.
17. Engelhardt, H. T., and Malloy, M.: 1982, 'Suicide and Assisting Suicide: A Critique of Legal Sanctions', *Southwestern Law Journal* **36**, 1003–37.
18. Francis, L. P.: 1980, 'Assisting Suicide: A Problem for the Criminal Law', in M. P. Battin and D. J. Mayo (eds.), *Suicide: The Philosophical Issues*, St. Martin's Press, New York.
19. Francis, L. P.: 1985, 'Poverty, Age Discrimination, and Health Care', paper given at Symposium on Poverty, Law, and Economic Justice, Santa Clara; *Logos*.
20. Fried, C.: 1970, *An Anatomy of Values: Problems of Personal and Social Choice*, Harvard University Press, Cambridge, Mass.
21. Fuchs, V.: 1984, '"Though Much is Taken": Reflections on Aging Health, and Medical Care', *Milbank Memorial Fund Quarterly* **62**, 143–66.
22. Gavin, M. R., and Kataja, G.: 1987 'Self-Determination in Later Life', in this volume, pp. 129–136.
23. Jackson, D. L., and Youngner, S.: 1979, 'Patient Autonomy and "Death With Dignity": Some Clinical Caveats', *New England Journal of Medicine* **301**, 404–8.

24. Kapp, M. B.: 1982, 'Response to the Living Will Furor: Directives for Maximum Care', *American Journal of Medicine* **72**, 855–9.
25. Kastenbaum, R.: 1976, 'Suicide as the Preferred Way of Death', in E. S. Shneidman (ed.), *Suicidology: Contemporary Developments*, Grune & Stratton, New York.
26. Martin, R. M.: 1980 'Suicide and Self-Sacrifice', in M. P. Battin and D. J. Mayo (eds.), *Suicide: The Philosophical Issues*, St. Martin's Press, New York.
27. Lamm, R. D.: 1984, 'Long Time Dying', *The New Republic*, (August 27), 20–23.
28. Lubitz, J., and Prihoda, R.: 1982, 'The Use and Costs of Medicare Service in the Last 2 Years of Life', Mss. in circulation.
29. President's Commission for the Study of Ethical Problems in Medicine and Biomedical and Behavioral Research: 1983, *Securing Access to Health Care*, 3 vols, Government Printing Office, Washington, D.C.
30. Rawls, J.: 1971, *A Theory of Justice*, Harvard University Press, Cambridge, Mass.
31. Rowe, J. W.: 1985, 'Health Care of the Elderly', *New England Journal of Medicine* **312**, 827–835.
32. Schultz, J. H.: 1976, *The Economics of Aging*, Wadsworth, Belmont, Calif.
33. Torrey, B. B.: 1984, 'The Visible Costs of the Invisible Aged: The Fiscal Implications of the Growth in the Very Old', paper presented to the American Association for the Advancement of Science.
34. Windt, P. Y.: 1980, 'The Concept of Suicide', in M. P. Battin and D. J. Mayo (eds.), *Suicide: The Philosophical Issues*, St. Martin's, New York.
35. Zook, C. J., and Moore, F. D.: 1980, 'High-Cost Users of Medical Care', *New England Journal of Medicine* **302**, 996–1002.

JOSEPH M. HEALEY

ELDERLY DEPENDENCY AND AUTONOMY: COMMENTS ON THE ESSAYS OF DUBLER, BATTIN, KATAJA AND GAVIN

I

The essays in this volume encourage the examination of our societal policy toward the elderly and they lead us to question what this policy has been, is, and ought to be. Now that aging and the interests of the elderly have become the focus of so much societal attention, it is important to move beyond the catch phrases and the clichés and to consider how the elderly, especially the dependent elderly, fare in our society. In their contributions, Molly Gavin, Gayle Kataja, Nancy Dubler, and Margaret Battin explore health care decision making by and for the dependent elderly and raise a series of critical questions:

(1) How should such decisions be made?
(2) Who should make such decisions?
(3) What range of options should be available?
(4) When and on what basis, should these decisions be subject to third-party control or review?

The three contributions under discussion serve to stimulate reflection on these issues and to provide suggestions about how these areas of emerging public policy should be addressed. They assist our search for balance and fairness in caring for the elderly, helping to establish the boundaries of compassionate care while avoiding both indifferent abandonment of those in need, as well as the paternalistic dominance of those vulnerable to mistreatment by others. A desirable public policy requires the inclusion of three key elements: (1) a realistic, accurate understanding of aging and of the perspective of the elderly person; (2) a candid assessment of the criteria and the procedures for justifiable societal intervention in the lives of the dependent elderly; (3) an honest acknowledgment of our ability and our willingness to provide resources for the care of the elderly.

191

Stuart F. Spicker, Stanley R. Ingman, and Ian R. Lawson (eds.),
Ethical Dimensions of Geriatric Care, 191–194.
© *1987 by D. Reidel Publishing Company.*

Each of the three essays addresses aspects of the (above noticed) four questions and of the three key elements of a desirable public policy. For example, Gavin and Kataja use their experience with Connecticut Community Care, Inc. (a statewide private non-profit agency), as a model for interdisciplinary, team management of the health needs of older adults. Such a team provides assessment of the patient, the coordination of services, benefits, and entitlements, and the monitoring of services.

The major focus of their essay is on two cases which illustrate dilemmas that confront the dependent elderly and those who provide care to them. Each case involves the wish of a dependent elderly patient to return home rather than to receive care at a nursing home. In each case, the self-determination of the patient is counterposed with the medical, ethical, legal and economic issues involved in respecting the individual's right to choose treatment *and* the treatment setting. The authors use the cases as a vehicle for raising the underlying issues concerning who should make such decisions and which standards should be used. They help us to appreciate in concrete terms both the perspective of the elderly person and the process through which societal intervention occurs.

In Professor Dubler's view, it should come as no surprise that autonomy, long a problematic issue in the doctor-patient relationship, should emerge as a major issue in the relation between the health practitioner and the dependent elderly patient. She describes the vulnerability of the dependent elderly who may be unable to effectuate their preferences without the assistance of others and who therefore require the cooperation of others. She provides an extended review of the concepts of autonomy and competence, emphasizing their role as the basis for respecting the wishes of the dependent elderly. Of special interest is her careful analysis of the influence of societal context on these concepts, reminding us that they have evolved in response to changing social needs and social settings. They are societal artifacts arising out of power relationships rather than neutral structures whose meanings are self-evident. She objects to the prevailing inadequate understanding of the aging process which leads some to equate aging with incompetence and to fail to respect the liberty rights of the elderly. Her review of conservatorship and guardianship proceedings reflects her concern that such procedures have a strong potential for discriminating against the elderly.

Dubler also examines several suggestions for resolving conflicts between the right of a patient with compromised abilities to make decisions, and the obligation of a responsible agency to protect that person or to protect itself from negative effects of those decisions. In the end, she concurs with Gavin and Kataja in supporting interdisciplinary consultation which would permit evaluation of available options, the patient's wishes, and, when necessary, court intervention. In this way, the boundaries of legitimate state intervention would be clarified.

Professor Battin's essay is a call for a realistic description of the aging process and of society's ability and willingness (or, perhaps more pointedly, inability and unwillingness) to devote adequate resources to the health needs and desires of the dependent elderly. Her essay examines two important issues: (1) the acceptability of suicide as a legitimate option for the dependent elderly patient; and (2) the impact of an age-based resource rationing scheme with regard to the dependent elderly. Her essay describes the disenfranchisement of the elderly which occurs in our society. She asks us to consider the social acceptance of planned suicide as an alternative to the denial of medical treatment to the elderly, which would result from the age-based rationing of health resources. In her conclusion, she reminds us that even if it is possible to justify planned suicide as an alternative to abandoning the elderly, both options may represent less desirable alternatives to a policy in which full medical treatment is made available to the elderly who wish it. We must recognize that the assumption of scarcity of health care resources which gives rise to either approach may not be accurate. The scarcities existing in medical resources are largely artifacts that could conceivably be reexamined and eliminated by a redistribution of resources. If this were desirable, the need for such approaches might be obviated.

II

In one sense, the issues raised by the contributors should provide no greater difficulties in cases involving the elderly than they should in cases involving any adult. Our legal system presumes that an adult is competent to make decisions without assistance or supervision, and such interventions are generally limited to cases where the person's decision-making capacity is compromised or the person's conduct poses a danger to others [1]. In the health care process, the pervasive influence of benevolent paternalism has been challenged and an alternative model that emphasizes shared decision making has achieved wide acceptance

[2]. Within this context, the elderly person should be presumed competent and should not be subject to intervention unless the general, applicable criteria are satisfied. The contaminating influence of "ageism" can be reduced by incorporating into our consideration of policy recommendations the three elements described above. First, there must be an understanding of aging and of the perspective of the elderly person which avoids both an unduly romanticized view of aging and an overly pessimistic view disguised as realism. A balanced view is needed. Second, the criteria and procedures for justifiable societal intervention must be articulated and assessed, and the inherent bias equating aging with disability must be confronted. Finally, our ability, willingness, or lack thereof to provide resources for the care of the elderly must be honestly acknowledged. Such efforts will contribute to our ability to respect the autonomy of the elderly person and avoid both abandonment and unwarranted paternalism.

School of Medicine
University of Connecticut
Farmington, Connecticut

BIBLIOGRAPHY

1. Annas, G. J. and Densberger, J. E.: 1984, 'Competence to Refuse Medical Treatment: Autonomy *vs.* Paternalism', *The University of Toledo Law Review* **15**, 561–596.
2. President's Commission for the Study of Ethical Problems in Medicine and Biomedical and Behavioral Research: 1982, *Making Health Care Decisions: The Ethical and Legal Implications of Informed Consent in the Patient-Practitioner Relationship*, Vol. 1: Report, U.S. Government Printing Office, Washington, D.C., pp. 36–39.

SECTION IV

JUSTICE IN THE PROVISION OF MEDICAL CARE FOR THE AGED

NORMAN DANIELS

EQUAL OPPORTUNITY, JUSTICE, AND HEALTH CARE
FOR THE ELDERLY: A PRUDENTIAL ACCOUNT

1. MOTHER AND DAUGHTER

About three years ago, my 91-year-old great-aunt was removed from the nursing home where she had been for several years and placed in intensive care at a major hospital. This was the last of several acute episodes suffered by a woman with significant mental impairment and physical disability. Prior to being placed in the nursing home, like many partially disabled elderly, she had found it difficult to get adequate home care and difficult to get an appropriate nursing home placement. Only in acute episodes did our health care system respond with extensive and expensive services. When I phoned her daughter, I was assured that the doctors were 'trying everything.' Our conversation proceeded something like this:

N: I know this may sound terrible, but should you let them try everything?

D: What do you mean? I can't let her suffer! We have to do something.

N: We cannot let her suffer, but are we prolonging her suffering? Have the doctors discussed what to do if she takes a bad turn – has a cardiac arrest?

D: No, they just said they're trying everything. But it's terrible to see her hooked up to all those tubes, and she doesn't even recognize me anymore. You know, I was the only one she recognized most of the time.

N: Yes, I know it has been very hard on her and on you. Perhaps you should talk to the doctors about letting her go in peace?

D: What are you asking me to do! It's my mother! I have to help her.

N: I don't want to press you, but, if you were in her condition and had lived a long, rich life, would you want your daughter to do for you what you are doing for her? Would you want to be treated that way?

D: God forbid. I'll never let her do it! When my time comes, I just want to go.

197

Stuart F. Spicker, Stanley R. Ingman, and Ian R. Lawson (eds.),
Ethical Dimensions of Geriatric Care, 197–221.
© *1987 by D. Reidel Publishing Company.*

N: Then why do you think you must continue aggressive treatment now?

D: I just have to – she's my mother! Wouldn't you do the same for yours?

I was struck by the contrast between my cousin's view of how she would want to be treated and how she felt compelled to treat her mother. Her own prudent view of how she would want health care resources allocated within her own life gave her no guidance in her decisions concerning her mother. Of course, how we would want to be treated is not always a good guide to how we ought to treat others, the Golden Rule notwithstanding. But I was impressed by the fact that we have created health care institutions which are, from a social perspective, imprudent. They lead to our being treated in ways we do not want to be treated, and they fail to meet other needs. My cousin, like her mother, will not get the home care services she will want and need if she becomes partially disabled, yet she will have to ward off intensive, acute care which she will not want but others may feel they are obliged to give her.

It seems to me that my cousin's prudent preferences about how she wants to be treated should have more to do with how her mother was treated than the circumstances – and the design of our health care institutions – allowed. I do not mean that my cousin's preferences for herself should directly determine how her mother is treated. Rather, my cousin's predicament, which we all share, raises a general question which we ought to answer namely: what is the relationship between the entitlements of the elderly in a just health care system and our prudential thinking about what we would like such a system to do for us?

In what follows I shall address this question, though I will be concerned with principles of justice that govern the design of our health care system, not with distributions to particular individuals, such as my cousin and great-aunt. I will argue that the socially prudent design of our health care institutions should be our guide to what justice requires in the treatment of the elderly. That prudence should guide justice in this way is not uncontroversial and has some controversial conclusions. My cousin's predicament hints at them, for generalizing her prudent preferences would lead her to treat her mother in ways she at first finds morally unacceptable. Moreover, I shall want to consider application of this approach to the controversial topic of rationing medical resources.

But before turning to the controversy, I want to motivate the approach and show its intuitive appeal.

2. SUBSIDIES AND SAVINGS

The central idea is intuitively easy to grasp: we all age, and as we do, we are affected by social institutions in different ways at different stages of our lives. This fact is elementary, but important. For example, as we age, we pass through social institutions responsible for distributing various social goods to us, in accordance with various distributive principles. If our needs vary at different stages of our lives, we presumably want these institutions – like health care systems – to be responsive to these changes. It would be prudent to design institutions so that our entitlements, our claims to goods and services, reflect these changes. In effect, this means we must be prudent about what goods and services we claim in certain stages of our lives so that we do not deprive ourselves of important benefits in other stages of our lives. In general, in budgeting prudently we may take from some parts of our lives to make our lives as a whole better. Specifically, this may take the form of 'saving' resources by deferring their use from one stage of life to another.

Budgeting in this way is unavoidable, unless we assume resources are infinite. This is a point worth emphasizing, since many people are reluctant to think about health care resources as limited. Two factors contribute to this reluctance. First, we see vast social resources used for purposes we think wasteful, misguided, or even destructive – such as building nuclear arsenals or 'star wars' weapons or subsidizing smoking. A just system, we tell ourselves, would not impose resource constraints on health care when it expends so much in these ways. Second, we see inefficiencies and inappropriate resource allocations within our health care system. Surely, we think, these resources could be reallocated in ways that dispel the apparent scarcity.

I think both of these objections to claims about scarcity have merit, but they should not lead us to ignore the reality of resource limitations. Even in a just and efficient system, there will be competition for resources among worthwhile projects as long as resources are finite. Such 'moderate scarcity' defines the conditions under which we need to worry about justice in the first place: justice is not a problem when manna falls from heaven, and just institutions cannot be sustained on a desert island. Moreover, persistent moderate scarcity is a practical reality, even if much of the scarcity is induced by injustice or inefficiency or imprudence elsewhere in the social system. So I propose here to treat scarcity as real, abstracting for most of this paper from the problem that some of it may be avoidable scarcity. Of course, my conclusions are qualified by this supposition.

Since they are the focus of my discussion, it will be useful to consider institutions which actually help us budget over our lifetimes. For example, our income support and health care institutions embody, in a very imperfect way, some features of prudent allocation over a lifetime. In the U.S., we defer income from our working lives to our post-work retirement period, which has generally been after age 65, by paying a payroll tax. This tax revenue from the employed is used to fund the benefits of currently retired workers. There are no 'vested' savings. Our payroll deductions do not form an annuity fund on which we draw in the future. But, if the social security system remains stable, young workers will be entitled to claim benefits when they age and retire. There is an intergenerational compact which has the effect of transferring resources from an individual's working years to his retirement years, in effect smoothing income over a lifetime in a way that is relatively well protected against inflation and the uncertainties of predicting individual lifespan (cf. [18]).

Though we do not think of it in the same way, our health care system has a similar 'saving' function. To see this function, we must think of the private and public insurance schemes as one set of institutions. We must add Medicare and Medicaid onto our system of largely employer-funded health care insurance. When we think of the system this way, we find that employed workers, young and middle-aged adults, pay the overwhelming share of all health care costs. Their average annual health care insurance premium – partly in the form of employer benefits, partly in the form of their own premiums, partly in the form of payroll taxes – far exceeds their actual health care costs. There is a transfer of health care resources from the young and especially to the old. Indeed, those over age 65 use health care dollars at roughly 3.5 times the rate of those under 65 (though it is a relatively small proportion of the elderly, the high cost users, who account for this statistical increase) (cf. [12]). In effect, society is community-rated as one risk pool, and the relatively healthy working adult population pays a premium that exceeds what is actuarially fair to it.

3. AGAINST SEEING DOUBLE

But the 'unfairness' of this inflated premium is only an illusion of perspective, a mirror in which we confuse others with ourselves. What is crucial about the health care system is that we pass through it as we age – the elementary fact I pointed to earlier. We may pay through the nose as working adults, but we free-load on it in our youth and old age. It is a system that transfers resources from stages of our life in which we have relatively little need for them into stages in which we do . If our perspective is that of the working adult who thinks only about his inputs and outputs in that stage of his life – which is, understandably, our everyday perspective – then the inflated premium seems unfair. If, however, we think of the system as a social savings scheme, and if we recall the claims the system gave us on health care we did not pay for in our youth, and if we anticipate benefits we will not (fully) pay for in our old age, then we see the cross-subsidy to others disappear. Instead we see that the inflated premium in our adult years is needed to pay for our needs in other stages of our lives. Of course, as I noted in the case of my great-aunt, our actual insurance scheme falls well short of being prudently designed in many of its features. Still, we all share in it as a scheme for reallocating health care resources from stages of our lives in which we have many resources and few needs into those stages in which we have fewer resources and greater needs.

I have just suggested that we gain insight into a difficult problem of justice – what is a just distribution of resources between different age groups? – by a shift of perspective. We must not see the problem as one between distinct groups – working adults who pay the high premiums and the frail elderly who draw claims on it. Rather, we must see that each group represents a stage of our lives and we must view the prudent allocation of resources among stages as our guide to justice between groups. My suggestion may seem a conjurer's trick. Now you see an inflated insurance premium, now you don't. Now you see the 'others,' the elderly or the young, benefiting from what is 'yours,' and now you don't. Instead, you only see yourself at other stages of your life, benefiting from your own (albeit imperfectly) prudent savings.

But this shift of perspective is no trick. It depends on that elementary fact I pointed to earlier, that we each age. Notice that age behaves differently from other traits which may serve to distinguish groups – like sex, race, religion, or class. Though we age, we do not change sex or race, and even religion and class are relatively fixed for an individual. This difference has profound implications for distributive justice.

Consider the effect of this difference between age and race on distributive schemes. From the perspective of stable institutions operating over time, differential treatment of people by age is differential treatment within a life and not between lives. Indeed, such differential treatment merely seems to appeal to a standard view of individual rational choice: it is rational and prudent that I take from one stage of my life to give to another, in order to make my life as a whole better. But the situation is quite different for race. Suppose we could increase aggregate well-being in a society by taking from one race and giving to another: perhaps some South African whites believe such is the case for apartheid. This defense may seem like a minor variation on the principle of individual rational choice: it says it is rational and morally unproblematic to take from some to make society 'as a whole' better off, as measured by aggregate happiness or well-being. Thus it treats taking from some to give to others on a moral par with taking from one stage of life to make the life as a whole better. But, in fact, this principle of social choice is radically different from the principle of individual rational choice: it involves crossing the boundaries between persons. Rawls [19] advances this point as a fundamental criticism of utilitarianism (cf. [16], [17], [2], [11]).

Distributive schemes which cross personal boundaries risk failing to respect the importance of persons. The difference between the race and age cases should now be clear. Distributive schemes that take age into account look like cases of crossing personal boundaries only if we adopt the perspective of a moment or 'time-slice' – the 'me' of the adult worker 'now' subsidizing health care for 'them,' the elderly. Once we take the perspective of institutions operating through time, the appearance of crossing boundaries between persons fades – the double-vision I complained of earlier disappears – and we are concerned primarily with distribution through the stages of a life. No comparable point is true for distributive schemes which treat race or sex differentially, even over time. The shift in perspective I urge is thus no conjurer's trick, but it is rooted in a real, distinctive, and morally important fact about age.

4. INSURANCE I WOULD LIKE TO BUY

I would like now to sharpen and refine the proposal I have been making about how to think about distributive justice between age groups. To do so, I would like to consider in more detail an argument I have sketched elsewhere about the moral permissibility of using age as a criterion for rationing life-extending resources – under certain conditions, of course. I will be particularly interested to contrast the kind of argument I am offering with other arguments, both pro and con my conclusion, for the contrast will reveal why the approach I am taking here avoids much that is controversial about rationing.

Before introducing the prudential argument in its full-dress form, it may help to give a more intuitive version of it. Suppose I know I have available to me a lifetime health care allocation, say in the form of an insurance benefit package. However, it is up to me to budget, once and for all, that allocation or benefit package so that it is used to meet my needs and preferences over my lifetime. How would it be rational for me to budget it, given all the uncertainties about my future health, wealth, and family situation? One plausible proposal might be for me to reserve certain life-extending technologies for my younger years, reasoning that doing so maximizes the chances of my living a normal life span. I might also use some of the resources 'saved' in this way to provide myself with more social support and home care services if I turn out to need them in my old age. I might reason that such services could vastly improve the quality of my years in old age and that such an improvement is worth the increased risk of a slightly shortened old age. I would then instruct, through my benefit package, the providers to treat me accordingly, that is, to appeal to an age criterion in their utilization decisions concerning me.

This insurance package is intended to resemble some features of the British National Health Service and to contrast with corresponding features of our own system. I have in mind the frequent reports that hemodialysis, for example, is not used on many of the elderly patients who receive it here: such resources are reserved for younger patients. At the same time, far more extensive home care services, alternatives to early – premature – institutionalization of the frail elderly are available. My understanding is that the British have tended to view elderly patients as 'medically unsuitable' for dialysis, though experience in the U.S. would tend to counter this claim, which begins to seem a disguise for rationing by age rather than medical condition. Since the British have often been criticized for this 'unequal' treatment of the elderly, the insurance package I just described, and the rationale I provided for it, are of some interest. For the rationale suggests that, if prudence governs the design of such insurance packages, and if we construe them as designs for a social, not private, insurance scheme, then we may be justifying a form of rationing by age that has received sharp moral criticism.

5. MY PRUDENCE, YOURS, AND OURS

The simple argument I have just sketched is open to an obvious objection. It involves an individual – Norman Daniels – reasoning about what he prefers and the prudent ways of satisfying those preferences. If all that is involved in such an argument is my giving explicit consent to my being treated differently at different points in my life, then that is one, relatively uncontroversial, thing. But other individuals may have different preferences and may want insurance schemes which differ, indeed which reject differential treatment by age or which involve different uses of age criteria. We face, in other words, the objection that there is no one prudent scheme for all individuals, because what is prudent is determined by an individual's preferences, resources, and individual needs. It is for this reason, for example, that my cousin's prudential reasoning about her own treatment might not be a basis for deciding what she ought to do for her mother.

One response to this objection is an attempt to let us have our cake and eat it too. Suppose people could choose among several lifetime health insurance packages, each of which represented a different conception of prudent lifetime allocation. We might even imagine that our publicly subsidized health insurance schemes, such as Medicaid and Medicare, had alternative benefit schemes and people could choose which would apply to them. Then no individual's judgments of prudence would be imposed on anyone else. My cousin and her mother would each have chosen an insurance policy, and my cousin's preferences would not determine what ought to be done for her mother. My great-aunt's own prudential judgments would. Thus prudence would play a role in the design of the health care system, and yet individual variation in judgments about prudence would be respected.

This scheme sounds fine until we realize that we would have to bind people to the choices they would make at a certain very early point in their adult lives. If we allowed people to change insurance plans as they age, we would create conditions that assure market failure. As Bishop has pointed out, we would open the door to adverse selection and moral hazard, and we would provide little protection against inflation and the effects of technology change [1]. Indeed, that is presumably why we have no private market for such lifetime health care insurance that includes long term care. But if, to avoid these problems, we required a once-and-forever choice by young adults, we would be biasing the plans towards the interests and preferences of young adults. Since people's situations and preferences change, we would not have really acknowledged what individuals would view as prudent in the later stages of their lives. This method of respecting the differences between persons fails because it would have to give undue weight to the choices made at a particular early point in life.

This consideration, and others I cannot discuss here, suggest that we need to modify our standard model of prudential reasoning. That model takes the choices of a fully-informed, rational, economic agent as its starting point. Such an agent, for example, knows how old he is and has much information about other conditions affecting his choices, such as his medical history, likely family situation, and so on. But the choice made by this economic agent at a given point in his life captures what is prudent for him in other stages of life in a very imperfect fashion. So the model, once it is restricted to single choices, as it must be to prevent market failure, only has the appearance of respecting individual choice about what is prudent, for it will lead to allocations that people do not view as prudent from their fully-informed perspective later in their lives. For the design of cooperative social schemes, such as health care systems, we need a perspective that abstracts in a reasonable way from the full-blown rational consumer used by the economist, but which still permits some form of prudential reasoning about lifetime health care allocations. My suggestion is to substitute a hypothetical deliberator who makes prudent choices under certain information constraints, a 'veil of ignorance' familiar in recent work on the theory of justice, e.g. in Harsanyi ([13], [14]) and Rawls [19].

These hypothetical choosers stand in for us in deliberating about the principles that should govern the design of our institutions. For this hypothetical contract to have any justificatory force, however, we must agree that the constraints imposed in the choice situation are procedurally fair and are appropriate to the task at hand. One line of argument justifying a veil of ignorance that keeps choosers from knowing their age, so that they might enter the system they designed at any stage of their lives, is that it prevents the choosers from appealing to special interests that characterize one stage of life. Eliminating this source of possible bias, which we saw would be a serious defect of the insurance scheme noted earlier, would clearly be a consideration in its favor. The result would be fair to choosers because it would not incorporate a systematic bias, say one resulting from accidental demographic features of the population of choosers. A full defense of the constraints would involve showing that they are procedurally fair to the choosers because they reflect their fundamental moral status, for example as "free and equal" moral agents (cf. [19], [20], and [4]). Developing this defense would carry me well beyond the limits of my task here. It is not enough, for example, to suggest that constraints on knowledge of one's medical history and family situation seem to be but exaggerations of the considerable uncertainty we face outside the veil in planning health, family, and economic eventualities over a lifetime, but I can do no more here.

6. EQUAL OPPORTUNITY AND PRUDENT ALLOCATION

Before sketching the prudential argument which justifies rationing by age, I must say briefly what I think justice requires of health care systems in general, since the prudential argument must be compatible with more general requirements of justice. Elsewhere I have argued that meeting health care needs is of special moral importance because it promotes fair equality of opportunity ([6], [10]). It helps guarantee individuals a fair chance to enjoy the normal opportunity range for their society. Specifically, health care, which maintains and restores normal functioning, has the particular and limited effect of allowing individual to enjoy that portion of the normal range to which his skills and talents would ordinarily give him access, assuming that these too are not impaired by special social disadvantages. The *normal opportunity range* for a given society is the array of 'life plans' reasonable persons in it are likely to construct for themselves. An individual's fair share of the normal range is the array of life plans he may reasonably choose, given his talents and skills. Disease and disability shrink that share from what is fair; health care protects it. Protecting an individual's fair share of the normal opportunity range is what the fair equality of opportunity principle requires health care systems to do.

One conclusion that follows from this general account is that we should use impairment of the normal opportunity range as a fairly crude measure of the relative importance of health care needs at the macro level. We need to refine the notion of normal opportunity range for its use in prudential reasoning about distribution between age groups. Life plans, we might note, clearly have stages, which reflect important divisions in the life cycle. Without meaning to suggest a particular set of divisions as a framework, it is easy to observe that lives have phases in which different general goals and tasks are central: nurturing and training in childhood and youth, pursuit of career and family in adult years, and the completion of life projects in later years. Of course what is reasonable to include in a life plan for a stage of one's life not only reflects facts about one's own talents and skills, tastes and preferences, but also depends in part on social policy and other important facts about the society.

The suggestion I want to introduce is that the prudent design of the institutions that affect us over the different stages of our lives requires reference to the notion of an *age-relative normal opportunity range*. Specifically, prudent deliberation about the design of such institutions, carried out with the degree of abstraction from individual perspective I argued earlier was appropriate to the task, would attempt to assure individuals a fair chance at enjoyment of the normal opportunity range for each life stage. We can provide a rationale for the suggestion that age-relative opportunity range should act as a constraint on the prudential reasoning of our deliberators.

Consider the perspective of designers of a health care system who are under an appropriate veil of ignorance. It keeps them from knowing their individual health status, their conception of the good (that is, their individual goals and fundamental interests by reference to which we define what is good for them), their age, income, and other important facts about themselves as individuals. (Remember, I have not fully justified selecting these constraints on knowledge, though there is discussion of similar constraints elsewhere in the literature on justice.) At the same time it lets them know important facts about the disease/age profile for their society, its technological level, and even that longevity has been increasing, largely as a result of other features of social policy. One feature of their problem emerges as critical: in choosing principles for institutions that defer the use of resources, *they must assume their own life-span will be normal*. Since they cannot appeal to any very special conception of the good, which might lead them to discount the importance of their projects or plans at a certain stage of their lives, they must treat these stages as of comparable importance. Here they are simply in compliance with Sidgwick's account of rationality: each moment of life is equally valuable and must not be discounted merely because it comes at one point in our lives rather than another. (I ignore problems with the Sidgwickian view [11], [16], [17]).

From their perspective, prudent deliberators will not know just what their situation is, what preferences or projects they might have at a given stage of their lives. Still, they do know that they will have a conception of the good and that it will define what is meaningful for them in their lives. But then it is especially important for them to make sure social arrangements give them a fair chance to enjoy the normal range of opportunities open to them at any stage of life. This protection of the range of opportunities they enjoy is doubly important because they know they may want to revise their life plans; consequently they have a fundamental interest in guaranteeing themselves the opportunity to pursue such revisions. But impairments of normal species functioning clearly restrict the portion of the normal opportunity range open to an individual at any stage of his life. Consequently, health care services should be rationed throughout a life in a way that respects the importance of age-relative opportunity range. In effect, the prudential task of the deliberators is constrained by the equal opportunity principle which governs health care in general and by other more general principles of justice.

7. PRUDENT RATIONING

Consider two implications of this view for the design of health care systems, keeping in mind that these systems operate through time on all stages of one's life. How would prudent deliberators view the importance of various personal care and social support services for the partially disabled as compared to personal medical services? From the perspective of these deliberators, both types of care would have the same rationale and same general importance. Personal medical services restore normal functioning and thus have great impact on an individual's share of the normal opportunity range. But so too do personal care and support services for the partially disabled and frail elderly. They compensate for losses of normal functioning in ways that enhance individual opportunity.

The frequency of partial disability increases for the elderly, especially for those over age 75. These disabilities are in general not life-threatening, and people usually live for many years with them. But they can have a dramatic impact on an individual's opportunity to carry out otherwise reasonable parts of his life plan and 'quality of life' may be sharply reduced – that is, if there are no personal care and social support services that promote independent living. It is not prudent to design a system that ignores these health care needs, since they affect such a substantial portion of the later stages of life, and which pays attention only to acute crises. This form of imprudence involves either not transferring enough resources to the later stages of life or transferring resources in the wrong form, as for instance claims on acute 'services instead of personal care and social support services.

A major criticism of the U.S. health care system, that it encourages premature and inappropriate institutionalization of the elderly, should be assessed in this light. The issue becomes not just one of costs and the relative cost-effectiveness of institutionalization vs. home care. Rather, opportunity range for many disabled persons will be enhanced if they are helped to function normally outside institutions. They will have more opportunity to complete projects and pursue relationships of great importance to them, or even to modify the remaining stage of their life plans within a greater range of options. Often this issue is discussed in terms of the loss of dignity and self-respect that accompanies premature institutionalization or inappropriate levels of care. My suggestion is that the underlying issue is loss of opportunity range, which obviously has an effect on autonomy, dignity, and self-respect. Viewed in this light, the British health care system, in which extensive home care services exist, far more respects the importance of normal opportunity range for the elderly than does our system. They put their resources into improving opportunity range for the substantial number of elderly disabled over significant periods of the late stages of life. We put our resources into marginally extending life when it is threatened in old age by acute episodes. Their approach is more prudent because it better protects age-relative opportunity range than ours.

I come now to the controversy I promised earlier, the prudential argument that justifies rationing by age. Under certain resource and information constraints, I shall claim, prudent deliberators would prefer a distributive scheme which improved their chances of reaching a normal life-span to one which gave them a reduced chance of reaching a normal life-span but a greater chance to live an extended span once the normal span is reached. Before considering this claim, it will be important to note a relationship between measures affecting longevity and rates of saving.

If lifetime earnings are held constant, but life-span is extended, then the rate at which resources must be transferred from early stages of life to late stages must be increased. We must take more from our young and middle years to finance our late ones. Of course, if we remain productive longer, then the savings rate may not have to be increased: but that is because lifetime earnings are not held constant. For example, if we extend average life-span but allow people to continue work through their young-old years, then savings may not have to be increased. Wherever life-span is increased during non-productive years, however, because we extend the period of old-old age or have more people reach it, then we must save at a greater rate in our productive years. Here, of course, prudence is at work: it would be imprudent not to provide resources for the late stages of life when we have reasonably good chances of having to live through them. Similarly, where increased longevity is primarily achieved by reducing early death, for example by measures which reduce infant or young adult mortality, the increased productivity that results will (we may suppose) roughly counterbalance the need to save more. Where increased longevity results from marginally extending the lives of the very old, especially those unable to work, then savings will have to be increased.

Under some resource constraints, the increased rate of savings needed to provide prudently for a life-span extended beyond the normal range will have serious negative effects on early stages of life. We can imagine constraints operating in the following way: providing very expensive or very scarce life-extending services to those who have reached normal life-span can be accomplished only by reducing access by the young to those resources. Saving these resources by giving ourselves claim to them in our old age is possible only if we give ourselves reduced access to them in earlier stages of life. A central effect of this form of saving is that we increase our chance of living a longer-than-normal life span at the cost of reducing our chances of reaching a normal life span. We see a version of this effect in recent health care policy. Infant mortality rates in Massachusetts have soared since 1980, which coincides with drastic cutbacks in prenatal care, especially for the poor [4]. In the meantime, we leave untouched the acute-care bias of the Medicare system which devotes vast hospital resources to marginally prolonging the life of the dying elderly, who are a substantial portion of "high-cost users" [23].

This point can be made more concrete by considering two rationing schemes. Scheme A (for Age rationing) involves a direct appeal to an age criterion: no one over age 70 or 75 – taken to represent normal life span – is eligible to receive any of several high-cost, life-extending technologies, e.g., dialysis, transplant surgery, or extensive by-pass surgery. Because age rationing reduces utilization of each technology, there are resources available for developing them all, though only for the young. Scheme L (for Lottery) rejects age rationing and allocates life-extending technology solely by medical need. As a result, it can either develop just one such major technology, say dialysis, making it available to anyone who needs it, or it can develop several technologies, but then ration them by lottery.

We may think of Scheme A as one that saves resources, that is, defers their use until later in life, at a lower rate than L. Scheme L takes more from earlier stages so that later ones may benefit. Specifically, Scheme L involves reducing the chance that the young will reach a normal life-span since they have reduced access to life-extending resources. Scheme L offers in return an increased chance to those who do reach normal life span of living a longer than normal span. For instance, though this is an extreme example, Scheme A might offer a 1.0 probability of reaching age 75 (and dying right away), and Scheme L might give an 0.5 probability of reaching 50 and on 0.5 probability of reaching 100. Both yield the same expected lifespan, but they do so differently. (Intuitions about science fiction cases are always of questionable utility, but for those who insist: Imagine there is a disease around which would kill everyone at age 50, but a drug is available in short supply. We can give a half-dose to everyone, and they will then live to 75 and die; or we can give a full dose to half the population by lottery, so that half die at 50 and half at 100.)

We may think of our prudent deliberators as being forced to choose between Schemes A and L (leave aside the science fiction example). I will argue that prudent deliberators would probably prefer an age-rationing scheme to a lottery, but the argument is complicated by the vague way in which I have described the choice situation, as anyone familiar with recent work in the theory of justice will have noted. I have left the description of the choice situation vague because defending a particular construction is difficult and, for our purposes here, digressive. But the price I pay for sticking to the point is that I must now consider

the way the argument might run under alternative constraints. Specifically, we must consider two alternative choice rules that might be
invoked to govern prudential reasoning about Schemes A and L.

One choice rule is the "maximin" rule (maximize the minimum),
which tells us to make the worst off outcome as good as it is possible to
make it, which might mean to make it as unlikely to happen as possible.
Maximin is appropriate to governing rational choices when real uncertainty, not just risk or probability, is a feature of the choice situation.
That is, it is appropriate when we cannot invoke likelihoods of outcomes, or even reasonably assume outcomes equi-probable and assign
numbers to them. Some might claim maximin is also the appropriate
rule when the worst outcomes are so grave that they cannot merely be
weighed against better outcomes.

I have not described the choice situation as one in which the maximin
rule is clearly the appropriate one. But if it is the correct rule, then I
think it is possible to show that Scheme A would be preferable to L.
Even someone who might defend L because he thought the life of the
wise, revered elder was the best thing he could aim for, and who would
not mind taking chances to reach the Golden Age (if he could calculate
the chances), would have to agree that the worst outcome for him would
be dying young. If we can assign no probabilities in our reasoning, for
example, if we can assign no probabilities to whether we are likely to die
young or live to the Golden Age, we are constrained by the maximin
rule to minimize the likelihood of the worst outcome. This would force
us to choose Scheme A, which has that effect.

I have not insisted on the maximin rule because I am not sure I want
to make the "veil of ignorance" surrounding our prudent deliberators so
thick that they cannot make the relevant likelihood estimates of longevity
under the two schemes. After all, I want our deliberators to know
enough about their social system so that they can make prudent judgments about the design of its health care system. As a result, they may
have to know something about its specific demography and the way it is
affected by alternative arrangements of the health care system. But this
suggests that we might adopt a more common rule of choice which
instructs prudent deliberators to maximize their expected net benefit or
payoff in choice situations. This Standard Rule, I shall call it, requires
that they take into account not only the value of a payoff, but its
likelihood or probability, and maximize the product of the two.

How would Schemes A and L fare under the Standard Rule? Suppose, for the moment, that we take as the payoff the number of years lived. The Standard Rule tells us to maximize expected life-span. If our choice between Schemes A and L is a choice between schemes which give equivalent expected life-spans – e.g., the choice between a 1.0 probability of living to (and only to) 75 under Scheme A and a 0.5 probability of living (only) until 50 and 0.5 probability of living only until 100 – then the Standard Rule instructs the prudent deliberators to be indifferent. If there is no better scheme, each is prudent and both are acceptable. This result is itself interesting. It says that age-rationing cannot in general be ruled out on the grounds that it is imprudent, which means it cannot be ruled out under the conditions I have argued are appropriate for deciding what is just between age groups.

The argument I have just sketched for the Standard Rule seems too abstract, even for the choice situation I have described, in which parties do not have knowledge of their conceptions of the good. Even without knowing the details of one's conception of the good, we do know enough about the frequencies of disease and disability age to know that years late in life, say after age 75, are far more likely than earlier years to be endured with some forms of impairment. This suggests that it would be imprudent to count the expected payoff of years late in life quite as highly as the expected payoff of years more likely to be free of physical and mental impairment. To be sure, many people enjoy years late in life relatively free of impairment – I am not drawing on a stereotype that all the old are frail and sick – and there is no suggestion here that age by itself gives us any basis for judging the value of these years to be less. Moreover, many people with impairments would admit to being no less happy than other people without impairments. Some people are happy and cope well, though others do not. But the prudent deliberators are making estimates of expected payoffs and have to take frequencies of disability and disease into account, and this means they may have to discount the expected payoff of late years accordingly. Specifically, this would incline them to view the choice between Schemes A and L, with the hypothetical expected payoffs assigned in the example, as misleading, and Scheme A would again seem the more prudent.

We might think about the choice between Scheme A and L under the Standard rule in a slightly different way, which may be an alternative to, or supplement to, the above argument. Under some plans of life, the contribution of the last years to the overall meaningfulness of life might be very great. But such Golden Age plans are probably atypical. Most people are well aware of their mortality and construct plans in which the tasks and rewards of early and middle years are integral to their success. For them, late years can be wonderful, but they are gravy to the meat and potatoes of the rest of life. Without making the judgment that one plan of life is better than another or even, by itself, less prudent, deliberators· familiar with their society and culture, but unaware of which conception of the good is theirs, might estimate it more likely they will have typical plans than Golden Age plans. They might then select Scheme A over L because they want to increase their chances that they live through the middle stages of their lives, for that is what will most insure success of their probable life-plan. Notice that if we want to block completely this type of probabilistic reasoning, then we start to push ourselves into a description of the choice problem that makes the maximin rule seem more appropriate than the Standard Rule, and then the deliberators would choose Scheme A in any case.

My conclusion from these admittedly sketchy versions of the prudential argument is that there are conditions under which a health care system that rationed life-extending resources by age would be the prudent choice and therefore the choice that constituted a just or fair distribution of resources between age groups. Since the argument I have offered leads to such a controversial conclusion, and since it is a complex argument with many presuppositions, I would like to distinguish it explicitly from some kinds of arguments – for and against similar conclusions – with which it may be confused.

8. WHAT THE PRUDENTIAL ARGUMENT IS NOT

The prudential argument I have offered focuses on rationing by age, but it has broader implications, specifically for discussions about the relative merits of extending biological lifespan or 'squaring the curve' (cf. [21] and the essays in [22]). The topic has thus attracted some discussion from diverse viewpoints. I want to be careful to distinguish the argument I offer from others, both pro and con my conclusion, for I think what is distinctive about my argument avoids much that is objectionable in these other arguments. In the course of these remarks, I will perforce reply to some objections to my argument.

Consider first an anti-rationing argument that appeals to equality. An age-rationing scheme will not treat the 75-year-old the same way it treats a 55-year-old, even though there may be no difference in their medical condition or need. But this is unequal treatment of like cases, for age is not a 'morally relevant' consideration. So any scheme that rations by age is morally objectionable.

The scheme I have justified by my prudential argument is not open to this objection, at least if we treat the notion of equality carefully and we observe that the schemes we choose operate over a lifetime. Specifically, the 75-year-old and the 55-year-old will not be treated differently over the course of their lives. Before each is 75, he will be entitled to the life-extending treatment in question; afterwards, he will not. Thus treating someone now 75 differently from someone now 55 does not constitute unequal treatment of the whole person over his lifetime. Moreover, it is question-begging to say that age is a morally irrelevant feature of persons. From the perspective of prudential planning over a lifetime, which is the appropriate perspective to adopt, it is not irrelevant. From this perspective, as I noted earlier, age must be clearly distinguished from race or sex. We age, but we do not change race or sex.

Consider now a pro-age-rationing argument that appeals to equality or fairness in a different way. We should give priority to the young in rationing life-extending services, so the argument goes, because the old have already had a chance to live more years and it is only 'fair' to the young to give them an equal chance.[1] This appeal to intuitions about fairness is not persuasive. Does it matter to our intuitions whether the old have already made claims on comparable resources to extend their lives, or is the occasion of competition with the young their first such claim? What if the young person has already received a lot of help, but the old person none? Is it still fair to the old person to deny him help now? When do we say the young person has used up his 'equal chance' to live to an old age?

I am not sure where such intuitions lead us or whether they are to be trusted. My prudential argument is different in that it appeals to no prior intuitions about fairness, except perhaps those which suggest the choice situation is procedurally fair to the prudent deliberators. The fairness of the outcome – what is just is what prudence recommends under these choice conditions – is derived from the fairness of the choice situation and the fact that it is reasonable for us to let these deliberators decide for us (cf. [19] and [20]).

One type of argument against age-rationing appears to rest on a fundamental claim about the equal value of life at any age and seems to deny part of the argument I offer above. Thus some insist that age-rationing is wrong because 'life at any age is worth living' or 'life at any age is of equal worth to life at any other.' This claim about the value of life has already been addressed in my prudential argument. From the perspective of prudent designers of a health care scheme using the Standard Rule, the most the equal worth claim can imply is that they should choose schemes that maximize expected lifespan. By itself, this does not decide the question for or against age-rationing. In particular, deliberators would have to consider both age-rationing and lottery schemes equally prudent if they had the same (maximizing) effect on expected life-span. They would then have to be indifferent between them, which means the equal value assumption does not of necessity lead to opposition to age-rationing. Moreover, if a lottery scheme failed to maximize life-expectancy as well as an age-rationing scheme, then the commitment to 'life at any age is worth living' would in fact commit one to choosing the age-rationing scheme. The assertion of equal worth should not be understood only from the perspective of the elderly person who knows he has already lived many years and wants more. It should be interpreted from the perspective of the prudent deliberator seeking a system that maximizes expected life-span, and then age-rationing is not ruled out.

Sometimes this 'equal worth' argument is coupled with the complaint that no one other than the person whose life is in question should judge its worth. This complaint seems to strengthen the anti-rationing argument because we may suspect that the rationing is being advocated by the young, who have an interest in it, and therefore any judgment that seems to deny equal worth is morally suspect. But it is an important feature of the prudential argument I offer that it involves no judgments by one person about the value or worth of another's life. Instead, it involves persons making judgments for themselves about benefits to themselves at different stages of their lives.

In this regard, the prudential argument works differently from those who argue in the opposite direction, for age-rationing, by claiming that ('their') elderly lives are not worth what ('our') younger ones are. For example, something like the latter judgment is embedded in certain methods for 'pricing' life, e.g., almost definitely in an earning-streams approach and probably in some willingness-to-pay methods. Someone might urge that it was more cost-beneficial or cost-effective to reserve certain resources for the young rather than the old because of the difference in 'quality life years' saved. There is a better return on the social investment of resources if the young get treated and the old do not. In the prudential argument I offered earlier, where reference is made to the increased frequency of impairments in later life or to the likelihood that an earlier year will be more important to a plan of life than a late year, the argument may sound similar. But the prudential argument involves reasoning about the expected benefit of an extra year at two points in one life – the life of the prudent deliberator. This judgment is not one 'we' make about 'them' but one whose consequences we must live with in each stage of our lives. The prudent deliberator is drawing the conclusion that his own life is less important to extend at one stage than it is at another – and that is the kind of judgment we do in fact make when we prudently plan our actual lives.

Though the prudential argument, especially under the Standard Rule, has some resemblance to a more general form of utilitarian reasoning, it plays a much more restricted role in my overall account of just health care and justice more generally. I advance the prudential perspective to address a particular and distinctive problem of distributive justice, justice between age groups, and the rationale for using that perspective turns crucially on the elementary fact I noted earlier, that we age. I am not urging this particular form of prudential reasoning as an account of just health care or justice in general. Indeed, I import into the prudential perspective, as a constraint on it, reference to my more general account of just health care, the fair equality of opportunity account. But it is easy to lose sight of this constraint in the argument about age-rationing of life-extending resources because life-extension by itself seems to have a comparable effect on age-relative opportunity range at any stage of life. Consequently, impairment of age-relative normal opportunity range would not decide the particular rationing question we were discussing, even if it does have a bearing on other allocation issues between age groups.

9. PRUDENCE AND REFORM

The last qualification, that the prudential perspective is only part of a more general account of just institutions and not such an account itself, points to some important limitations on the argument about rationing by age. It is easy to misconstrue and to misapply my argument. The argument does not, in general, sanction rationing by age. Such justification is possible only under very special circumstances. First, it is crucial that the appeal to an age criterion is part of the design of a basic institution that distributes resources over the lifetime of the individuals it affects. Nothing in the argument offered here justifies piecemeal use of age criteria in various individual or group settings – e.g., by some hospitals or physicians, or in any way that is not part of an overall prudent allocation.

Second, the argument should not be taken as a hasty endorsement of age rationing as a convenient "cost-constraining" device in the context of current debates about our health care system. Not only is such an application not likely to be part of the design of our basic health care institutions, construed as a savings scheme, but many of the assumptions about resource scarcity which might make rationing by age prudent in some circumstances are controversial in the context of this public policy debate. This does not mean that my prudential perspective cannot be suggestive of some reforms we might aim for, and I will mention some briefly in a moment.

Finally, it is important to see that my argument is part of an ideal theory of justice, in which we can assume general compliance with principles of justice that govern other aspects of our basic social institutions. The argument does not readily or easily extend to non-ideal contexts, in which no such compliance with general principles of justice obtains. Thus it would be wrong to say that my argument actually justifies the British system of rationing dialysis by age (assuming that is their practice). At most my argument shows that such rationing can under some circumstances be part of a just institution, that it is not always morally objectionable in the way that sex or race rationing would be. The argument shows the conditions that would have to obtain for such rationing to be just.

My main purpose in this paper has been to urge that we approach the problem of distributing resources between age groups from the perspective of people prudently allocating resources over their life times. I focused on the argument for rationing by age not because this is a pressing issue of public policy for us now, but because it was a useful device for making us take seriously the more general prudential perspective. That perspective does suggest some reforms of our health care system that would move it in the direction of a more prudent one, reforms far less controversial than rationing by age. One reform, the expansion of our long-term-care system to include personal care and social support services, I have already pointed to. Were our system made more prudent in that direction, my poor great-aunt would have had a far better old-old age and my cousin, now in her 70's, could look forward to the kind of help she is most likely to need. I also think my account would give priority to improved long-term-care over further development and use of exotic life-extending services for the old-old, but I have not directly argued this point; I have only argued the stronger point about rationing by age.

We can, however, make important changes in our practice of prolonging the last few months of life without resorting to an actual rationing policy, and my prudential account would certainly encourage such changes. Much could be accomplished if people were educated to think prudently and prospectively about the kind of health care they want as they age through the last stages of life. I believe that much that is imprudent in the design of the system derives from that fact that our insurance schemes – public and private – have had very little input from prospective patients concerning the benefit package over long time periods. If people were compelled to think about these issues prospectively, perhaps because they had strong incentives for doing so, many would not want the drastic measures now taken on their behalf. We should not take as our norm, or as a model of what people want, the allocation decisions that result from family panic, physician enthusiasm, and lack of clear planning and instructions from the patient.

Any scheme that incorporated prudent reforms would be undermined if we did nothing to change the thinking and the incentives, both perceived moral imperatives and legal sanctions, of providers. Much must be done to educate physicians about the 'gatekeeping' decisions that they make. The intention should not be to persuade physicians to adopt an under-the-table form of rationing by age. Such rationing, if we are driven to it, must be the result of a social decision that forms a framework for allocation decisions binding on providers. Just rationing should be our social decision, not their individual one (cf. [5], [7], and [10]). Rather, the intention should be to force careful thinking about the implications of what may seem to be "merely" medical decisions and challenges. Similarly, we must examine very carefully the structure of cost-containment schemes, such as DRG's, to see what incentives they actually – and not just theoretically – give providers for treatment of the elderly, and we must assess the implications of these incentives for the prudent design of the system. (Unfortunately, and imprudently, funding for research to assess these financial reforms of the system has been reduced, so it will be difficult to determine actual effects.)

In all of these reform settings, however, we enter the political debate as actors with very particular interests. It is thus easy to forget that we are concerned with systems that treat us over the course of our lives. We focus on what will happen next and not on what has been done over time, including what has been done to us. In this political world, it is hard to think of the elderly as later stages of our lives, or the young as earlier ones. These stages become whole people who do not perceive their interests as coincident with ours. The mirror of the urgent present deceives us into thinking only about 'them' and 'us' and not about ourselves over time. The moral point of view for solving distributive problems about age, the prudential perspective, requires a serious effort at distancing ourselves from our current, immediate situation. It requires breaking the illusion of the mirror. This distancing might seem hypothetical and psychologically unreal, but for one important fact. We need not adopt the perspective of other people, only one that recognizes that these institutions affect us over our whole lives and that we must budget within them accordingly. At least for these savings institutions, we should do unto others as we would have done to ourselves.

ACKNOWLEDGEMENT

Some of the central ideas for this paper derive from my earlier work ([8], [9], and [10]), but I wish to thank Hugo Bedau and especially Joshua Cohen for helpful discussion of the prudential argument for age-rationing, which has been developed here. Research for this paper was funded by the National Endowment for the Humanities and the Retirement Research Foundation.

Tufts University
Boston, Massachusetts

NOTE

[1] Veatch [22] offers a variant on this argument in which the principle requiring the equal chance is itself derived from a contractarian argument intended to yield a general principle of justice for health care. I implicitly reject Veatch's argument in my [3], [6], and [10].

BIBLIOGRAPHY

1. Bishop, C.: 1981, 'A Compulsory National Long-term-care Insurance Program , in J. J. Callahan and S. S. Wallack (eds.), *Reforming the Long-term-care System*, D. C. Heath, Lexington, Mass.
2. Daniels, N.: 1979, 'Moral Theory and the Plasticity of Persons', *Monist* **62**(3), 265–87.
3. Daniels, N.: 1979, 'Rights to Health Care: Programmatic Worries', *Journal of Medicine and Philosophy* **4**(2), 174–191.
4. Daniels, N.: 1980, 'Reflective Equilibrium and Archimedean Points', *Canadian Journal of Philosophy* **10**(1), 83–103.
5. Daniels, N.: 1981, 'Cost-Effectiveness and Patient Welfare', in M. Basson (ed.), *Ethics, Humanism and Medicine*, Alan R. Liss, New York, pp. 159–170.
6. Daniels, N.: 1981, 'Health Care Needs and Distributive Justice', *Philosophy and Public Affairs* **10**(2), 146–179.
7. Daniels, N.: 1981, 'What is the Obligation of the Medical Profession in the Distribution of Health Care?', *Social Science and Medicine* **15**, 129–133.
8. Daniels, N.: 1982, 'Am I My Parents' Keeper?', *Midwest Studies in Philosophy* **7**, 517–540.
9. Daniels, N.: 1983, 'Justice Between Age Groups: Am I My Parents' Keeper?', *Milbank Memorial Fund Quarterly/Health and Society* **61**(3), 489–522.
10. Daniels, N.: 1985, *Just Health Care*, Cambridge Studies in Philosophy and Health Policy, Cambridge University Press, Cambridge, England.
11. Daniels, N.: 1987, *Am I My Parents' Keeper?: An Essay on Justice Between the Young and the Old*, Oxford University Press, New York (in press).
12. Gibson, R. M., and Fisher, C. R.: 1979, 'Age Differences in Health Care Spending, Fiscal Year 1977', *Social Security Bulletin* **42**(1), 3–16.
13. Harsanyi, J. C.: 1955, 'Cardinal Welfare, Individualistic Ethics and Interpersonal Comparisons of Utility', *Journal of Political Economy* **63**, 309–321.
14. Harsanyi, J. C.: 1976, *Essays in Ethics, Social Behaviour, and Scientific Explanation*, D. Reidel Publishing Co., Dordrecht, Netherlands.
15. Knox, R.: 1984, 'Fund Cuts are Linked to Infant Death Rise', *Boston Globe* **225**(145) (May 24), 1, 20.
16. Parfit, D.: 1973, 'Later Selves and Moral Principles', in A. Montefiore (ed.), *Philosophy and Personal Relations*, Routledge and Kegan Paul, London, England.
17. Parfit, D.: 1984, *Reasons and Persons*, Oxford University Press, Oxford, England.
18. Parsons, D. O. and Munro, D. R.: 1978, 'Intergenerational Transfers in Social Security', in M. Moskin (ed.), *The Crisis in Social Security: Problems and Prospects*, Institute for Contemporary Studies, San Francisco, Calif.
19. Rawls, J.: 1971, *A Theory of Justice*, Harvard University Press, Cambridge, Mass.
20. Rawls, J.: 1980, 'Kantian Constructivism in Moral Theory', *Journal of Philosophy* **77**(9), 515–572.
21. Vaupal, J. W.: 1976, 'Early Death: An American Tragedy', *Law and Contemporary Problems* **40**(4), 73–121.
22. Veatch, R. M. (ed.): 1979, *Life Span: Values and Life-Extending Technologies*, Harper and Row, New York.
23. Zook, D. J. and Moore, F. D.: 1980, 'High-Cost Users of Medical Care', *New England Journal of Medicine* **302**(18), 996–1002.

STANLEY R. INGMAN, DEREK GILL, AND JAMES CAMPBELL

ESRD AND THE ELDERLY:
CROSS-NATIONAL PERSPECTIVE ON DISTRIBUTIVE
JUSTICE

As the proportion of aged and our clinical ability to treat the ill aged increase in the U.S., the news media, health care planners and politicians are asking whether we can afford ever-increasing increments of medical care. Within the various sub-worlds of medical treatment, debates about cost vs. benefits are also more frequent. Questions of rationing vs. open access are being raised by various authorities. To try to clarify the issues in terms of both national health policy as well as the allocation of scarce resources, we believe that a review of geriatric and renal care in the US, the UK and Switzerland is useful. Our basic concern is to explore the consequences of collectivist contrasted with individualist approaches in the allocation of resources to care for the elderly and persons with End Stage Renal Disease (ESRD). What we mean by 'individualist' compared with 'collectivist' orientations toward the allocation of resources will become increasingly clear in what follows.

Our starting point, however, is that a social system that operates with an individualist orientation is one that emphasizes the duties and responsibilities of individuals to provide for their medical care; a social system that operates with a collectivist orientation, on the other hand, is one that emphasizes the duties and responsibilities of the society at large to provide medical care to all.

We assume that societal policy and structures set conditions or broadly delineate what is likely to happen in one-to-one interchanges in important ways. Clearly, decisions affecting the allocation of resources at the national, regional, or local level determine indirectly, but very significantly, the nature of physician/patient relationships and may even have an impact on the choice of various treatment modalities for specific illness conditions.

There are several reasons why we chose the renal dialysis and transplant program as an example of medical technologies that impact on the

223

Stuart F. Spicker, Stanley R. Ingman, and Ian R. Lawson (eds.).
Ethical Dimensions of Geriatric Care, 223–262.
© *1987 by D. Reidel Publishing Company.*

aged. The initial consequence of the 1972 U.S. decision to fund ESRD treatment modalities out of Medicare had the effect of increasing the total budgetary needs of this form of medical insurance, since large proportions of early ESRD patients were pre-retirees. As the average age of ESRD patients rose, however, the proportion of Medicare funds absorbed by ESRD therapies continued to increase. Thus, a treatment modality aimed at a certain catastrophic illness, which initially encompassed only young and middle-aged adults with primary kidney failure, soon was needed to treat all persons so affected. Emphasis upon ESRD is appropriate and unique since it is the only disease category in the U.S. singled out for special federal funding. Moreover, the relatively high costs involved in ESRD programs have encouraged other medical care systems, particularly the British, to restrict access to this treatment modality.

We have chosen to compare the United Kingdom, Switzerland, and the U.S. for a number of reasons. The United Kingdom, with its National Health Service (NHS), has tried to design a publicly owned system to address equity of health care issues. The U.S. medical care system has a public sector that relies on financial and regulatory mechanisms to address questions of equity and distributive justice, whereas the British relies on administrative fiat. Switzerland, more similar to the U.S. in terms of political orientation and organization, depends on a mix of private insurance and mostly public ownership of the large hospitals and nursing homes to support and control its health care delivery within its twenty-six Cantons. The growth of private hospitals is currently on the rise, with some American hospital corporations moving into Switzerland. By selecting these societies we are able to evaluate various approaches to macro-allocation in three industrial societies and analyze how system characteristics influence the interaction between ethical dilemmas stemming from certain clinical situations and the way they are dealt with.

GERIATRIC AND RENAL CARE IN THREE NATIONS

We plan to summarize the evolution of geriatric and renal care in Britain, Switzerland, and the U.S. This will involve descriptions of geriatrics and renal care as separate elements of each nation's medical care systems, but there is an inevitable overlap since in all these countries it is the treatment or non-treatment of elderly with ESRD that gives rise to controversy. We will characterize, in broad terms, the evolution of these programs, their current status, weaknesses and strengths.

The British Geriatric Care System

Care of the elderly in Britain is embedded in the National Health Service and the nature of the geriatric services clearly reflects the broader aspects of the umbrella organization. The concept of a *national* health service clearly implies the ideal of delivery of medical and health care services uniformly to all members of society. The second implication of a national health service relates to the principle of equity and further implies that medical care services should be provided in such a way that no one is barred from access to them. In general, this means the delivery of services at zero cost at point of delivery. If medical services are not "free" at point of delivery, those least able or unable to bear the cost of such services are denied equity [22].

These principles of equity and uniformity in the distribution of medical services are only possible under the over-arching principle of public accountability. The principle that distinguishes a NHS from all other forms of medical and health care delivery is that it is a delivery system accountable, through the body politic, to the population it services. In practice, the degree of public accountability of a national health service may be limited, as is the case in Great Britain. Nevertheless, once the principles underlying a national health service have been accepted, the functional and professional autonomy [19] of the medical care industry is limited by the government.

The principles of a health service – universality, equity and accountability – are in a sense sequels to a political principle developed much further in British and Continental political systems than has been the case in the U.S.: greater emphasis on the rights and responsibilites of the collectivity than on those of the individual. The nations of Europe have seen the emergence of powerful collectivist, Socialist, and Communist political parties, which have had an obvious impact on the political ideologies and social realities of European society [29]. The "establishment", the dominant traditional élites in these societies, have had to respond to these collectivist pressures stemming from left-wing ideologies and to come to terms with a shift towards a collectivist ethic and away from the historically dominant individualistic ideologies of the eighteenth and nineteenth centuries. The development of national health services and other reformist medical care delivery systems are the social structural consequences of the gradual emergence of a collectivist ideology. The British NHS is one illustration of this phenomenon.

Geriatric care in Britain goes far to realizing the principles of equity, uniformity, and public accountability, but, as is well known, severe gaps in the extent and provision of services persist. Nevertheless, most components of what would now be recognized as 'state of the art' geriatric care and services are present, although many of the care components are either inadequately developed or bereft of the necessary financial support to provide equal and uniform coverage.

Geriatric care and services in Britain include geriatric hospital specialists (geriatric consultants and other hospital consultants), general practitioners, community based nurses, health visitors, social workers and services, long term accommodations, day hospitals, meals on wheels, old people's clubs and other social activities, sheltered accommodations and other types of housing support, home helps, and information services such as the Citizens' Advice Bureau [9]. Medical care needs are delivered first through General Practitioners, all but 10% of whom are in group practices or in the 1500-odd community health centers. General practitioners cooperate with health visitors, district nurses (and some with attached social workers) who provide primary care for all patients registered with their practice. GP's are encouraged to retain age/sex registers and their attached health visitors visit elderly persons in their own homes to screen for emergent social and medical problems. The primary care medical team, together with social workers when available, can coordinate medical and social services to maintain elderly persons in independent existence in their own homes through applying for the assistance of home helps, delivery of meals on wheels, home nursing care through the district nursing service, positions in day care attendance centers, and day care hospitals. If independent existence is no longer possible, general practitioners can refer patients to hospital geriatricians for specialist assessment and admission to a long term care facility, the majority of which are owned and operated by local authorities ([7], pp. 54–75).

Unfortunately, virtually all of these services are inadequate, not in conceptualization, but in extent and availability. There are insufficient health visitors, district nurses, home helps, sheltered and other purpose built and serviced housing, and long-term care facilities. Since the inception of the welfare state in the late nineteenth and early twentieth century and its expansion phase in late WWII and the immediate post war years, virtually all branches of the medical and social care systems have received limited and inadequate costs and capital and development funds. Even in the NHS hospitals, that sector of the medical care system which has received the major proportion of public funds since the inception of the service in 1948 is often outdated, capital equipment is in short supply, and nursing and medical services are understaffed. Indeed, it is increasingly becoming clear through the work of O'Connor [47], Offe ([48], [49]), Gough [24], and Navarro [45] that a welfare state in a mixed capitalist economy is virtually condemned to inadequate funding for both operating and capital funds.

ESRD in Britain

A brief outline of British geriatric services indicates the lack of association between the well-conceived welfare state philosophy and actual service provision. The treatment of persons with ESRD reflects the same discrepancy between ideals and reality. As Cooper points out, the Department of Health and Social Security intervened directly in 1971 when the issue of financing dialysis services surfaced ([10], p. 92). The department issued directives to Hospital Boards indicating the sum of money to be spent on dialysis care, and how and where the money should be spent. With the changes in administrative procedures introduced by the White Paper, and reforms of 1974, plus subsequent further changes in administrative procedures, local hospital boards have had additional flexibility in treating ESRD cases. Hence, while patients are not normally eligible for dialysis care if the onset of the disease occurs after the age of 60, some hospitals and regions do accept older patients. Nevertheless, the rate of treatment[1] [47] for ESRD in Britain in 1978 was 127 per million (all forms of dialysis and transplant), a rate which is less than that achieved by most other European societies at a similar stage of industrial development ([52], p. 92). Moreover, of those on dialysis, 65% dialyze at home, a procedure which economizes on formal medical care expense by imposing costs on informal care providers ([54], p. 997).

All treatment modalities for British patients are housed in and funded through the hospitals of the NHS. There are no free-standing and only three privately funded ESRD services in the U.K., the latter virtually restricted to the treatment of patients from overseas. Parsons ([50], p. 14) has concluded that ". . . significant curtailment [of ESRD treatment] is being implemented in the United Kingdom by limiting the dialysis population," and Gabriel ([20], p. 36) has stated that "the annual number of untreated patients is of the order of 1400 to 1700. This estimate includes potential patients up to the age of 70 years." Thus, it is evident that the principles of equity and uniformity are breeched in the treatment of persons with ESRD, particularly elderly patients.

Swiss Geriatric Care

The twenty-six Cantons[2] of Switzerland with their 6.3 million people have relatively independent health care policies. Some 50% of medical expenses are covered directly in Switzerland through a combination of communal, cantonal, and federal funding sources.

In 1890, the federal constitution encouraged citizens to purchase medical care insurance and establish regulations for both the commercial and non-profit insurance industries ([43], p. 76). As a consequence of the high standard of living, approximately 95% of the population is covered, and access to hospital and physician care is open to virtually all citizens. Reimbursement is on a fee-for-service basis, the mechanism of payment varying by insurance provider, with some insurance companies and sickness funds paying physicians and hospital charges directly, and others reimbursing patients' bills. Switzerland, like all other western societies, has been faced with the problem of rising costs for medical care as well as budget deficits. In 1979, the federal government, alarmed at its increasing fiscal responsibility for the support of medical care, placed a ceiling on the amounts it was prepared to make available to the cantons to support the care of the elderly and other needy groups. Subsequently, the cantons and the communes have had to increase their subsidies to the medical care sector to maintain service provision.

Major hospitals and the majority of nursing homes are owned and operated by government or by private non-profit foundations. Salaried physicians, as in Britain, work in and control the hospitals. Private fee-for-service general internists and specialists deliver ambulatory medical care to the aged and non-aged. This latter group of physicians has limited hospital privileges, or involvement.

Insurance schemes, whether non-profit or for-profit, reimburse more for technical and diagnostic procedures than for socio-clinical consultations, health education and preventive measures. In Switzerland, this has led to "diagnostic centers" being established in physician offices, e.g., small laboratories, radiological and endoscopic equipment may be housed on the premises. In 1975, the bulk of medical expenditure covered hospitals (45%) and private medicine (32%) ([43], p. 52). Home health care and social work services are under-developed both by British as well as other European standards. Swiss citizens are paying ever-increasing medical care insurance premiums. Incidentally, Swiss industry does not pay and has never paid a significant proportion of health insurance premiums – currently only 1% of the total premium ([62], p. 76). Public meetings are currently being held to discuss and debate who or what is to blame for this cost explosion in medical care costs [62].

Overall data on the Swiss medical care system are hard to locate because of the decentralized government. A less satisfactory but adequate alternative is to focus on one Canton's health care system for the aged. The Canton of Geneva is a good choice, for several reasons. Informally, the Canton was requested by the confederation to undertake an R&D function in developing an improved geriatric care system. Geneva, along with the Cantons of Berne, Zurich, Lausanne, Basel, and Vaud are richer than the twenty other more rural, non-industrial cantons of Switzerland.

By 1983, for a population of 340 000 with 13% aged 65 or older, there existed the following partially integrated components in a geriatric care system: (1) two new geriatric hospitals (80 and 276 beds), (2) two day hospitals and five day-care centers for the disabled and frail elderly, (3) some 1800 sheltered housing units with nurses on call, (4) a strong geriatric community nursing and home-health aide service (with 80 teams), (5) a geriatric ambulatory center (with five teams consisting of physicians, a physical therapist, a social worker and a psychologist to visit nursing homes and the residences of home bound persons), (6) a geriatric-liaison service at the 2000 bed university-cantonal hospital (created to improve inpatient care and discharge planning, as well as to prevent unnecessary hospital admissions) and, since 1965, (7) a geriatric psychiatric service for persons 65 and older (300 out of 600 beds at the university psychiatry hospital). Recently, a chair in geriatric medicine was proposed at the University of Geneva School of Medicine [31].

Of the 58 nursing homes and boarding homes that exist in the Canton, the majority of beds (55%) and the larger facilities are government or foundation (non-profit) operated [32]. Many cantons in Switzerland have no for-profit nursing homes, in sharp contrast to the robust private sector in the U.S. where more than 75% are for-profit. A significant range of quality exists among nursing homes and some of the best facilities in the world are to be found in Geneva. Apparently geriatric care in Geneva is poorly balanced and favors acute medical care services to the relative neglect of home care. This neglect of home care facilities and services for the aged may reflect the fact that Swiss medicine, unlike the U.K., has a weak general practitioner service.

ESRD in Switzerland

Dialysis programs operate out of the public hospitals located in the twenty-six cantons, and for-profit dialysis centers are almost non-existent. Thirty-one ESRD treatment centers or "stations", mostly in cantonal hospitals, serve some 260 ESRD patients per million, the highest rate in Europe and twice as high as in the U.K. Only Japan and the U.S., whose rates approach 290 per million population, service more patients ([33], p. 19).

In 1975, ESRD stations regardless of size or location received a fee of 400 SF per session ($254 U.S.). In 1982–83, the umbrella reinsurance organization (CLM) increased the payment to 500 SF ($318 U.S.)[3] VESKA (the national hospital association) lobbied for the 1983 increase but with limited support from ESRD programs. Some station directors were reluctant to support the request for an increase since it defeated, they claimed, their own attempts to control costs through home care, Continuous Ambulatory Peritoneal Dialysis, or self-care units. This reluctance on the part of some directors was also linked to their fear that profit-dialysis programs would emerge in Switzerland. Clearly, reimbursement rates or charges for ESRD do not reflect treatment costs. Rather, the renal program indirectly subsidizes other hospital services at the discretion of hospital administration.

In the 1950s Switzerland established CLM, a company to reinsure 86% of health insurance companies against catastrophic illness or continuing disability. In 1966 CLM extended its coverage to renal dialysis and kidney transplants. From a study of the ESRD stations in the Canton of Schaffhausen and from interviews with station directors in other locations, the following conclusions were drawn: (1) ESRD treatment costs are routinely covered by insurance programs; (2) patients spend little time worrying about treatment costs or reimbursements; (3) out-of-pocket charges for ESRD or related medical care are very minor; and (4) the standard of living of ESRD patients is relatively unaffected by ESRD involvement.

In 1977, 10 out of 31 ESRD centers in Switzerland reported that they excluded persons 65 and older. Fourteen stations reported that they restricted access based on whether renal patients had diabetes ([52], p. 98). In 1983, station directors in Geneva, Lausanne, Fribourg and Schaffhausen said they excluded no one [33]. The average age of hospital dialysis (1977) patients was 51.3 years, similar to the U.S. figure of 52 years but significantly higher than the average 40.6 years in the U.K. ([52], p. 98).

In addition to formal selection policies, exclusions occur informally in all countries to varying and unknown degrees. For example, a medical director of a skilled nursing home in Schaffhausen flatly stated that he does not and would not refer residents for ESRD treatment. In both the U.K. and Switzerland, the idiosyncratic way in which physicians evaluate ESRD patients has been documented by Taylor [57] and Wauters et al. [63]. A list of 30 renal failure cases were submitted to 29 Swiss dialysis centers and total agreement was reached in only one case of rejection and one of acceptance. Wauters concluded that socio-cultural factors seem to be the most critical determinants of selection or rejection. Wauthers also submitted 40 case descriptions from the U.K. study to the same Swiss ESRD stations. Despite the fact that approximately the same proportion of cases were accepted or rejected by each country's physicians, consistency in individual cases was rarely achieved.

Home dialysis represents one-fifth of the total population, a little less than 30 per million (1980), whereas the British figure of 60% generates a rate of 32 patients per million. Home dialysis varies widely from station to station from Schaffhausen with no home dialysis [33] to Lausanne with 29% in their programs [64].

U.S. Geriatric Care

Before the advent of Medicare and Medicaid, no formal organization existed within medicine to provide services to the elderly. There still is no specialty in geriatric medicine, although groups within internal medicine, family medicine and to a lesser extent psychiatry have, since the early 1970s, begun to take a special interest in the elderly.

Pressure for some form of financial assistance began to accumulate in the 1960s when medicine (through the early stages of its scientific and technical expansions) began to provide increasingly effective, but also increasingly costly, treatment modalities for a whole range of illnesses and diseases. Those sectors of the population covered by insurance policies were partly insulated from the rising costs of the fee-for-service based medical care delivery system. Most medical insurance policies either ceased to provide coverage or provided only limited coverage at retirement. Consequently, many Americans, including middle and upper class groups, found themselves financially embarrassed when the need for increasing amounts of more costly medical care began to impinge on their life styles after retirement [41]. Medicare emerged in 1965 to ameliorate this situation. In addition, Medicaid, care for the medically indigent, was introduced in 1966. It soon became apparent that many elderly persons were also eligible for Medicaid.

With Medicare and Medicaid programs introduced in 1965–66, financial assistance for three sectors of the medical care industry was improved markedly: hospital medicine, physician services, and nursing home care. Home care, other non-institutional services, and long term rehabilitative treatment remained poorly covered. After persons reached the poverty level or below, long-term care in nursing homes was covered by Medicaid [4]. Nursing home use rose in the '60s and '70s, but facilities remained isolated from mainstream medical practice, and quality of care was deemed marginal [60].

Access to expensive medical technology for the aged is less restricted in the U.S. than in most industrial societies of the world. Access to non-institutional services is limited by various copayment requirements or other eligibility restrictions. Upgrading the quality of long-term care as well as achieving continuity of care have become two major foci of the geriatric reform movement in the U.S. A current and growing concern is the resolution of actual or potential conflicts between the availability and distribution of complex and expensive medical technologies and the preservation of life. These concerns, in turn, raise problems that create dilemmas around instigation and cessation of treatment modalities [59]. On occasion, this problem is no longer constrained technologically; rather, in some instances, biological functioning can be maintained for varying periods without apparent consciousness. Constraints now relate to cost/benefit and quality of life issues.

The major objective of Medicare was to lift financial concerns and constraints from retirees. In fact, elderly patients now pay more in real terms for medical care than they did in 1964. Among the elderly, out-of-pocket expenses rose steadily and in 1977 only 41% of their personal health care expenditure was covered by Medicare ([1], p. 24). Indeed, many private insurance companies now provide policies to cover the gap between Medicare benefits and medical costs. These circumstances are particularly noticeable in the nursing home sector where many incumbents soon "spend down" to the officially defined poverty level and become dependent on the Medicaid program. This issue is further complicated by the recently introduced regulations governing transfer of funds and property to relatives and eligibility requirements of Medicaid [18].

This fiscal plight of the elderly will worsen as a result of two recent changes in Medicare regulations introduced by the Reagan administration: (1) raising the deductible and copayment amounts for both Part A (hospital care) and Part B (out-patient care), and (2) shifting some inpatient services that were paid at 100% [under Part A], to reimbursable only under Part B at an 80% rate. The justification for these changes is that the elderly will now pay more for their medical care and also bear an increased financial penalty if they over-utilize services. Since no attempt is made to distinguish between "needed" and "unneeded" medical care, this argument seems obscure.

Fiscal constraints have also limited to 100 the number of hospices (of a total of 1000) who have applied for certification to take Medicare patients. The Medicare program pays a fixed daily rate which may not exceed a total of $6500 per patient for care throughout this period to death [11]. The National Hospice Organization explains the reluctance of its members to apply for certification on the likelihood that many patients, once accepted into the program, will require services and care that will often far exceed this sum. This policy is particularly ill-conceived since the alternative is to continue to treat the terminally ill in the far more expensive, traditional, acute care facilities. A more satisfactory alternative would be to integrate hospice and hospital treatment facilities so that flexibility and interchange of patients, as medically appropriate, could take place, thus maximizing the benefits to the terminally ill of both institutions. This is a further example of the kind of stop-gap measures forced on the U.S. health care system because of the overall lack of planning, coordination, and integration within the system.

It is too early to make an assessment of the on-going switch – from retrospective to prospective reimbursement policies of financing Medicare and Medicaid through the Diagnostic Related Groups (DRG) system – on the care of the elderly. Nevertheless, the underlying philosophies of the DRG based system are based on increased competition between hospitals, savings through increased managerial efficiency and replacement of costly therapeutic regimens with cheaper methods when these are available and bio-medically acceptable. Since we have shown that Medicare programs are already under-financed, it may be assumed that any further economies imposed on the geriatric sector can only affect, deleteriously, the quality and quantity of medical care for the elderly [61].

Of indirect, but no less significant concern to Medicare beneficiaries, are the more fundamental problems that harass the U.S. medical care system. The recent introduction of the DRG system and the changes in Medicare regulations are specifically designed to reduce expenditure on medical care costs, at a time when 11% of the U.S. population is without any form of health insurance [55]. Moreover, the true figure is clearly much higher since it does not include the unemployed, the underemployed (who are often without health insurance benefits) and the hidden unemployed (those unemployed who have given up seeking work).

Most insurance companies, including Blue Cross and Blue Shield, provide inadequate coverage for categories such as catastrophic illness and long term nursing home care. Ambulatory care is inadequately covered and domiciliary services of all kinds are under-developed. Moreover, the U.S. spends less on non-defense expenditure than other industrialized Western nations:

Sweden allocates 56.5%, West Germany 42.2% and Japan 28% of their G.D.P. in general government outlays for nondefense expenditure, compared with just 26% in the US. All these countries have larger social (including health) government expenditures than the US. Sweden spends 33.8%, West Germany 30.6%, and Japan 27% of its GNP on social welfare expenditures (which include unemployment benefits; workmen's compensation; social assistance to children, the handicapped and others; health care; and old age and disability insurance) compared with only 14% in the U.S. ([46], p. 322).

Rather than attempting to reduce health care expenditure, the Federal Government should be doing all in its power to increase the resources allocated to the health and welfare sectors as well as to review the overall structure of the health care system with a view to improving its distributive efficiency.

ESRD in the U.S.

The first artificial kidney was developed during the Second World War in Holland. In the 1950s, dialysis was reserved mostly for acute renal failure in both Europe and the U.S. In the 1960s, with improved access procedure, chronic dialysis became feasible. Large numbers of dialysers were developed and many technical problems were solved. However, funds remained a problem and dialysis was limited to those individuals who were judged to be valuable to society. Potential candidates were reviewed by selection committees consisting of physicians, sometimes other health professionals, and community representatives. Selection criteria were not standardized, although a patient's age was often used [35]. For example, the average age of dialysis patients prior to 1972 was approximately 35, but by 1982 the average age had increased to over 55.

Between 1962 and 1972, the Veterans Administration, the National Institutes of Health, and the Public Health Service established service and research ESRD programs throughout the U.S. However, maldistribution of dialysis machines persisted and rationing of existing resources was standard practice in the period during which the new technology was undergoing development. After 1972, when passage of an amendment to the Social Security Bill extended Medicare coverage to all kidney patients,[4] "treatment for all" became the mode of operation and selection committees for renal failure patients were disbanded.

Why did the U.S. government enter the arena through the Medicare system? Kutner (1982) argues that the "major impetus was a desire to avoid the moral dilemma of indirectly deciding who could live in a country which considered itself to have almost unlimited resources" ([36], p. 53).

Tables I and II provide basic data on ESRD treatments in 1980–83 ([37], [21]).

In 1972, about 40 patients per million population were being treated by long term hemodialysis care and 40% of dialysis treatments was home based ([37], p. 76). Now, a little over 10 years later, patients on dialysis treatment exceed 250 per million population and less than 19% are on home dialysis. Over the same period the average age of dialysis patients has increased. By 1995, the Health Care Financing Administration estimates there will be 90 000 dialysis and transplant patients in the U.S. at a total cost to Medicare funds of $3 billion ([16], p. 71). This will represent a treatment rate of approximately 300/million population, a rate achieved by a funding policy that currently places few fiscal

TABLE I

End stage renal disease: Basic data for the U.S.

Category	Number	Date
No. of patients	61 445	1980
Payments	$1 207 600 000	1980
Transplant centers	156	July '81
Dialysis facilities & centers	1093	July '81
Haemodialysis home training centers	536	January '81
CAPD training centers	382	January '81
Kidney transplants	4697	1980

[Used with permission.]

TABLE II

Number of certified end-stage renal providers or facilities, 1980 and 1983 [21].

	Number 1980	Number 1983	Percent 1980	Percent 1983
All facilities	1073	1309	100.0	100.0
Proprietary	323	504	30.1	38.5
Hospital based	23	14	2.1	1.1
Hospitals	23	14	2.1	1.1
Hospital satellites	0	0	0.0	0.0
Freestanding	300	490	28.0	37.4
Non-Profit	750	805	69.9	61.5
Hospital based	620	668	57.8	51.0
Hospital	593	630	55.3	48.7
Hospital satellites	27	30	2.5	2.3
Freestanding	130	137	12.1	10.3

[Used with permission.]

constraints on patients or providers. Before going on to discuss the factors associated with policies affecting ESRD in each of the three nations, we must explore possible epidemiological causes of variations in rates.

Variations in Rates of Treatment of ESRD: Bio-Medical Factors

Three possible categories of explanation immediately come to mind, although the categories may not be truly separable, either empirically or theoretically: race, age, and disease prevalence. The census of the 1978 ESRD Medicare dialysis population revealed that Blacks constituted 27.1% of those being treated (Blacks constitute 11.7% of the US population). The prevalence estimate for Blacks derived from these figures is 482/million compared with 172/million for whites. Most European nations have relatively small proportions of their population of negroid origin and the higher rate of ESRD in the U.S. may reflect this circumstance. Prottas concludes that 49% of the difference observed in rates between the U.S. experience and those European countries with relatively high treatment rates may be due to the differences in the negroid/white proportions ([52], p. 95). However, even this conclusion must be treated cautiously since numerous hypotheses could be proposed to interpret this association. For example, race, as we know, correlates with social class in the U.S. and lower class whites also have higher rates of ESRD. However, few types of renal failures seem to be caused by preventable diseases.

Since age is positively related to the need for ESRD treatment, we might therefore expect nations with the highest proportion of elderly persons to have a larger at-risk population and hence higher ESRD rates. In fact, the U.S. has a younger population than most European societies, yet has the highest ESRD treatment rate in Europe and North America.

Geographic variation in the incidence of precursor diseases, such as hypertension, diabetes mellitus, and nephropathies as measured by mortality rates, showed no statistically significant variation within European countries or between Europe and the U.S. It has already been shown that blacks are subject to higher rates of ESRD than whites and it is surprising not to see this fact emerging in the above comparison. Nevertheless, without much more detailed information, including morbidity data, it would be hazardous to reject the null hypothesis.

GERIATRIC CARE, ESRD AND DISTRIBUTIVE JUSTICE

1. Great Britain

Service provision for the elderly and for persons suffering from ESRD invokes dilemmas centering on the allocation of resources. This is particularly problematic for both groups, since one element of apparent rationality, cost/benefit comparisons, seems less relevant to retired persons and to a disease where in some treatment modalities debility is so severe that employment opportunities become restricted or absent. This observation, of course, is one of degree and is itself the product of socio-historical forces that determine definitions of work and non-work. The point here is to show that the value positions affecting both micro- and macroallocation decisions are a complex mix of social, economic, and political forces. These forces are permeated by a given society's general beliefs concerning the value to be placed on an individual both as an individual and as a member of a collectivity or group. Thus, at the same point in time, the elderly person or the patient with renal failure is a psychological, economic, and sociological entity – an individual, a member of the elderly or renal failure patient group, and a member of the larger social structure. These points are all obvious, but they do serve to emphasize that, given finite resources, when a decision is taken to allocate goods or services to A(1) all other A(2-n) are affected; hence the problem of public accountability.

Now it is clear why the concept of public accountability is one of the basic components of a national health service and is a significant part of the British NHS. Although the process of allocating medical resources may not reach an exact standard of public accountability, the aggregate of individual decisions is subject to public accountability through the process of Parliamentary government. The introduction of the NHS in 1948 was not accompanied by a statement of objectives, but it is now generally accepted that the basic principles of the NHS include the precepts of equity and uniformity. The Office of Health Economics, for example, stated in its recent document: "Understanding the NHS in the 1980s," [the NHS] "does not seek to question the fundamental principle that the NHS exists to provide comprehensive health care for everyone in the population, regardless of their wealth or social status" ([56]. p. 3). Thus it does seem reasonable to evaluate the provision of geriatric care and services for ESRD patients under the NHS in accordance with the principles of equity and uniformity.

It seems clear that both in geriatric and ESRD care the NHS falls short of its declared objectives. In effect, ESRD care is severely ra- tioned since insufficient funds are provided to the Regions to treat all the ESRD cases occurring in their areas. Moreover, the relative poverty of the Regions varies enormously so that the severity of the rationing mechanisms differs in different parts of the country. In 1976, the Report of the Resource Allocation Working Party offered some suggestions to achieve a more equitable distribution of resources across the Regions but, as Heller has shown [28], opposition from élite groups within the medical profession has been considerable and the amount of funds set aside for reallocation is so small that years must pass before any significant improvement will occur.

Geriatric care is also bedeviled by a lack of resources, but the short-falls are less dramatic. A shortage of long-term care facilities is partly compensated for by exploiting women, usually daughters or daughters-in-law, who must care for their elderly in their own homes [15]. Nevertheless, the facilities to maintain elderly persons in their own homes in Regions with a tradition of cooperation between medical and social care can approach the ideal [34].

It could be argued that in circumstances that prevent all from being treated (ESRD) or where services are spread too thinly (geriatric care) other principles of justice might lead to a more reasonable distribution of scarce resources. Thus, instead of rejecting candidates for ESRD treatment on age alone, some assessment of the applicants 'social worth' or 'blameworthiness' in the etiology of his/her condition could be taken into account. Indeed, Carter-Jones has argued that such criteria are used in Britain today [8]. It would seem that such arguments are relatively simple to reject on practical psychological and political grounds. Physicians or other decision-makers who reject ESRD candidates on social criteria are simply assuaging their own uncomfortable mental state. Moreover, by attempting to use judgments based on the social worth of the potential patient, the decision-makers would be reducing the pressures on politicians, the group that must bear the ultimate responsibility for the lack of sufficient resource allocation.

It is important to note that this analysis of the British situation has generated a hierarchy of values. Given the objective of the NHS – to provide comprehensive health care for everyone in the population, regardless of their wealth or social status – the morally responsible actor must acknowledge the primacy of the principles of equity and uniformity. In practical terms this means political agitation for the allocation of more funds to ESRD services and geriatric care. Again the point is obvious but it does emphasize the responsibility of the ethicist to involve him or herself in political activity, a point which is too often overlooked.

2. The United States

Issues relating to distributive justice are much more complicated in the U.S. health care system. Medical care is funded through a complex mix of public and private sources and risk sharing is spread across individuals (out-of-pocket expenses, e.g., ambulatory care, copayments, and deductibles), the government, and private for-profit and non-profit insurance companies. Eleven percent of the United States population (perhaps considerably more) is without any form of medical insurance.

While Medicare is predominantly publicly funded, and Medicaid virtually completely so, the U.S. tax system is only mildly, if at all, progressive. Therefore, it is necessary to examine the tax mechanisms through which public expenditure on geriatric and ESRD services is generated if the discussion of distributive justice is to be taken further. Thus, most taxation on corporate income is passed on to consumers, basically workers and small businessmen, and is not borne by corporate owners. Gift and inheritance taxes are levied at a minimal rate and are easy to avoid. Commercial property taxes are shifted mainly to tenants, to consumers or back to workers. Payroll taxes are almost entirely shifted to the work force ([47], pp. 203–20). Excise and sales taxes are, of course, steeply regressive and families who earn less than the taxable ceiling still pay anything from 5–10% of their income in tax. Currently 45 states levy sales taxes. The social security tax is only slightly progressive and the progressive element closes at $43 800 income (until quite recently, the upper limit was $19 000 per annum). It is only in direct taxation, both Federal and State, that a progressive element is introduced, and even then compared with other Western societies the degree of incremental deduction from rising wages or salaries is much less than that of other tax systems in Western societies. Clearly the principle of distributive justice in the U.S. tax system is minimal.

The British taxation system, although much more progressive than either the Swiss or American systems, still includes elements favoring the wealthy, as is to be expected in a mixed economy. This is particularly true in periods, such as the mid 70s to the present in the United Kingdom, when the taxation system was and is being used to support and subsidize the private sector (Concorde, export credit guarantees, financing of arms sales, subsidies for research and development costs, etc.). The gradual development of the welfare state and its concomitant social expenditure had generated a small but significant redistributive effect in the total economy. For the classic 'average' family of two adults and two children the net effect of the tax structure and welfare state social expenditure was to reduce the range of income from 4.5:1 to 3.2:1 in 1973 ([24], p. 111). However, one of the first measures introduced by the Thatcher government was lower tax rates for middle and upper income groups; thus the range between upper and lower will now be even broader than in 1973. However, it is the comparative perspective that concerns us here, and one of Gough's conclusions seems appropriate:

Countries raising a greater share of finance from social security contributions (even if wage-related) or indirect taxes, without subsidies for housing or other necessities and without a national health service would, on the evidence of this study, redistribute income still less, if at all, between income groups ([24], p. 111).

The first two or three generations of Medicare and Medicaid recipients enjoyed a considerable amount of intragenerational redistribution of income. Current and subsequent generations of beneficiaries will obviously receive much less.

Above all, the U.S. tax system is unfair and finally becomes basically regressive through the notorious problems of evasion and distortion, in a word – loopholes. It is said that anyone with an income over $50 000 per annum and who pays any income taxes is either (1) intensely patriotic, (2) a fool, or (3) employs a lousy tax consultant.[5] Thus even the concept of distributive justice is inapplicable to the fiscal mechanisms through which Medicare and Medicaid are funded.

Geriatric care in the U.S., not surprisingly, reflects the complex mix of public and private funding service provision. The vast majority of nursing homes are operated on a commercial for-profit basis and partially financed through public funds.

In the intermediate and so-called skilled nursing homes, the vast majority of the staffs are nursing assistants with a leavening of licensed practical nurses and at least one registered nurse per shift in the skilled nursing facilities. The nursing assistants are paid at, or a little above, the minimum wage. These circumstances, often associated with poor quality premises and amenities, generate environments for nursing home residents more like medical warehouses than community homes for the elderly. Indeed, it is the predominance of the medical model – that all frail elderly persons are to be treated as un-well [12] – that justifies the organization of nursing homes along the lines of a hospital environment

[13]. The predominance of the medical model is further enhanced since the predominant mode of funding care of the elderly comes from the medical rather than the social care sector, the latter severely under-funded in the U.S. compared with European societies.

The fragmented nature of medical and social care for the elderly in the primary care sector is even more apparent. The fee-for-service concept that controls both the public and private sector of medical care is an incentive to lack of continuity in geriatric medicine, itself yet to achieve speciality status in U.S. medicine. The lack of coordination, planning, and integration in the social services sector is even more evident. Boone County in Missouri, for example, an administrative unit with approximately 90 000 population, has over 250 welfare and/or social service agencies and organizations, a situation that would be duplicated in almost any community in the U.S. [14].

Yet, from a taxation system that takes hardly any account of distributive justice and a funding system for medical care that is only partially publicly financed, persons with kidney failure are assured of treatment at considerably reduced personal expense. Some states also run Kidney Programs that provide additional assistance to providers and patients, but the latter have to meet eligibility requirements similar to Medicaid in many states. For many middle income patients and the 'near poor,' deductibles and copayments can generate considerable hardships [5]. Further, additional expenses are covered by the Social Security legislation under Titles II and XVI, which provide income support for the disabled. In 1979, 35 percent of the newly enrolled ESRD beneficiaries as classified as "ESRD disabled" [26]. Approximately 23 000 individuals with ESRD were receiving disability benefits at a cost of a further $100 to $250 million. Tables III, IV and V provide additional information on the costs of ESRD treatment [37].

Thus in 1980, 61 445 ESRD patients in all forms of treatment absorbed approximately $1.5 billion (Medicare and Titles II and XVI of Social Security), an average cost of $24 412 per patient. Although ESRD patients represent only 0.2% of the total Medicare population, they accounted for 5% of Medicare funds in 1979 [38]. Transplantation, from a fiscal point of view, is clearly the treatment of choice and, as we have already seen, steps were taken in 1978 to promote this procedure. Unfortunately, many ESRD patients are bio-medically unsuitable candidates for transplant surgery, and kidneys remain in short supply

TABLE III
Comparative annual costs 1981

Hemodialysis	$
Hospital	28 800
Center	24 100
Home	14 850
Peritoneal dialysis	
Continuous ambulatory peritoneal dialysis	18 300
Continuous cycling peritoneal dialysis	19 700
Transplant	10 000

Source: Adapted from the Kidney Dialysis Industry, 1981[6]
[Used with permission.]

TABLE IV
U.S. medicare transplantation costs & charges 1980–81[7]

	$
Average kidney acquisition charge	
Living related	7 581
Cadaveric	6 992
Average charge/transplant	20 156
Estimated cost/transplant	14 403

[Used with permission.]

TABLE V
Reimbursement sources for ESRD in Missouri, 1980–81

Total charges (25 centers)	$	%
Payments	26 244 787	100
Medicare	16 673 514	63
Medicaid	1 028 791	4
+Missouri Kidney Program	1 323 175	5
Other – insurance cost	4 186 126	16
++Unpaid	3 033 081	12

[Used with permission.]

despite attempts by political and medical authorities to encourage the harvesting of suitable kidneys from accident victims and others.

Although North Americans with ESRD have access to a wider range of treatment modalities than any other group of similar patients in the Western world, why was it that this particular group or collectivity was singled out for indemnification from the costs of their disease in a medical care system that emphasizes the duties, rights, and responsibilities of the individual? A partial explanation will be attempted ("partial" only because the categories of explanation would need extensive additional socio-historical research if complete support for the following hypotheses were to be achieved).

Attempts to introduce a collectivist orientation into American political and economic life have been remarkably unsuccessful. Trade unions represent only about 20% of the US labor force and Lens' history of the American Labor Movement is aptly entitled *The Labor Wars* [40]. By contrast, in Britain, France, the Low Countries and Scandinavia, a majority of the labor force is represented in trade unions. Moreover, these trade unions play a significant role in political as well as economic activity. The General Council of the British Trades Union Congress can exercise considerable power on behalf of its member organizations, whereas the AFL–CIO was powerless to prevent President Reagan from dismembering the air traffic controllers union.

The political components of the U.S. working class movement have fared even worse. Perhaps the most notable achievement of the early working class movement was the polling by Eugene Debs of 900 000 votes for the Presidency in 1920 when he was still in prison having been convicted under the war-time Espionage Act. In general, the period was characterized by the ruthless destruction of the political organizations of the working class by the Establishment through any measures, both legal and extra-legal, which proved effective. The constant harrying and eventual destruction of the Industrial Workers of the World was yet another example of the impossible circumstances under which communists and socialists tried to develop political as well as industrial representation for their memberships. The fate of the American Socialist Party was described by Milton Cantor in the following words:

The Socialist Party was also shaken by government attacks – by denial of seats to its elected representatives, deportation of its alien members, harassment of its leadership and its newspapers. In part, its weakening was due to the strength of vigilantism and, conversely, to the erosion of conventional democratic policies and politics ([6], p. 79).

In these circumstances it is hardly surprising that early attempts to introduce health insurance schemes into the U.S. met with overwhelming opposition. While some physicians expressed interest in the English experiment of 1911, the anti-European and anti-Socialist hysteria engendered by America's entry into World War I soon enabled the American Medical Association and the private insurance companies to label state health insurance programs as Bolshevik propaganda or Prussian Imperialism. After the Red Scare of 1919 and the Palmer raids of 1919–20, any attempt to advocate, let alone introduce, state health insurance schemes was doomed to failure [44]. Even today America has yet to endorse the principles of equity, uniformity, and public accountability in relation to the medical care system. More importantly, the overall American value system still supports an individualist as opposed to a collectivist morality.

Nevertheless, the U.S. Congress imposed a collectivist solution on the dilemma of high cost of ESRD treatment and the inability of the majority of individuals to pay – the state would undertake fiscal responsibility. The decision was, no doubt, made easier by the use of the dramaturgical method. The President of the National Association for Patients on Hemodialysis and Transplant was dialyzed on the floor of the House of Representatives. Paradoxically, the dramaturgical method, through focusing on the distress of an individual, forced the dilemma of a collectivity (all ESRD patients) on the attention of the Congress. In a sense, the distress of the President of the N.A.P.H.T., so evident at the level of an individual, could only be addressed by recourse to a resolution mechanism which responded to the needs of the collectivity. In other words, in this instance, the frailty of an individualistic morality that places responsibility on the individual for what becomes of him/her can only be dealt with or ameliorated by the collectivity undertaking some degree of responsibility or support for individual persons who find themselves in dire circumstances through no fault of their own. Successful lobbying activities by ESRD patient groups and the medical profession, plus press exposure of the uncomfortable position that nephrologists were forced to play when matching excess patient demand with inadequate treatment resources, also played a part. But above all, it would appear that the basic premise of distributive justice prevailed, i.e., that the costs of treating ESRD should be borne by the society as a whole.

Other possibilities more in line with the long tradition of resistance to collectivist pressures were theoretically available. Thus, legislation could have been introduced requiring profit and non-profit health insurance companies to cover ESRD, leaving Medicare and Medicaid to deal with the uninsured and uninsurable. A reinsurance procedure similar to the Swiss system could have been introduced. Hospitals could have been reimbursed directly instead of indirectly for ESRD services. None of these options was chosen, rather, P.L. 92–603 incorporated the collectivist solution. This decision, although enacted under a Republican President, may have owed much to the rather more liberal atmosphere that had been associated with the sixties, and the Kennedy and Johnson administrations which, as it were, "spilled over" into the early 1970s. Certainly the pros and cons of some form of a national health insurance scheme were still being debated and formed important planks in both the Ford and Carter campaigns, although the concept was soon dropped after the election.

Over the last decade in the U.S. virtually anyone who can benefit from ESRD treatment has been accepted into the program. Today the question posed is, are too many being treated? On occasion elderly persons are accepted for dialysis even when as candidates for transplant their overall life expectancy may be estimated in months rather than years. Most people would agree that the British age cut-off at 60–65 is too low, but might agree that some arbitrarily chosen but more 'reasonable' older age (such as average life expectancy of the individual's cohort, after which ESRD treatment would not be available) should be established. The management of scarce resources is basically a political problem and any real solution must have a political component. Surely it is not beyond the wit of society to generate a solution. Politicians are always reluctant to raise taxes, but what if a portion of an already existing tax were periodically earmarked for the treatment of all forms of catastrophic illness? Such a procedure might have the apparent and unfortunate consequence of encouraging "single issue" political controversies, but that might not altogether be a bad thing. It might not be a bad idea to require each politician or group of politicians who wish to propose, say, the purchase of an additional B-1 Bomber, to state from which proportion of which particular tax base the necessary funds should be extracted. Certainly such a procedure might inject much needed clarity and visibility into the process of government.

A further ethical dilemma, unique to the U.S. treatment system for ESRD, is the issue of profit. However, this issue need detain us but briefly since examination of Tables III and V will quickly expose the illogicality and immorality of the proponents of for-profit dialysis centers. The major for-profit company running dialysis centers is National Medical Care Inc. (NMC). From a $9 million volume in 1971, NMC reported $190 million dollars in revenue in 1979, with profits of $19 million. This resulted from treating 17% of the nation's dialysis patients in 1979. They also own Erika (dialysis supplies and equipment) and Life Chem (a clinical laboratory). Table III shows that hospital hemodialysis costs $4700 per annum more than center hemodialysis. NMC claims the savings result from the greater efficiency of their operation compared with hospital treatment facilities [38]. The efficiency argument is simply a smoke screen to hide the fact that the case mix is different in hospital and center dialysis and that hospital dialysis involves much higher overhead in terms of medical and nursing skills. Moreover, hospital dialysis teams are also expected to make some contribution to the total overhead of operating a hospital. These overheads include support for the 'uneconomic' sectors of hospital care, neonatal intensive care, pediatrics in general, uncollected bills, treating the medically indigent and, in the case of teaching hospitals, educational expenses. At least some of these costs are reflected in the final item of Table V.

Any six-year-old running a lemonade stand soon learns that a profit can only be turned by charging more for the product than the materials and the energy incorporated into the lemonade's production actually cost. Exactly the same process applies to NMC. Moreover, the $19 million in profits are achieved after paying exceedingly generous salaries to at least some of the staff. A nephrologist previously associated with an NMC unit in suburban Washington, D.C., said the director of that unit makes in the order of $400 000/year ([38], p. 381). Clearly, the efficiency argument is indeed a smoke screen and there is little to be said in support of a system that generates profits out of people's sickness, misery, and ill-health. The profit component becomes even more objectionable when it is realized that the excess of income over expenditure is largely publicly financed, and the elimination of profits would generate cost-savings that could be applied to other forms of needed social or medical care. In effect, profits are simply a form of wasteful expenditure.

3. Switzerland

In Switzerland, as in the U.S., issues pertaining to distributive justice and to equity, uniformity, and public accountability in relation to geriatric care and ESRD services are complex. Central government is relatively weak, with the Canton and to a lesser extent the Communes playing a much more significant role in the financing and operation of health care services. Consequently, local political power becomes an even more important consideration in Switzerland than is the case in either the U.S. or the U.K.

The values of Swiss society, in terms of its political traditions and institutions and social philosophy, are reflected in Rousseau's work. Switzerland's frequent recourse to the referendum perhaps comes closest to Rousseau's philosophy concerning relationships between the citizenry and government. In the latter half of the eighteenth century, city government might have been possible in a country of 726 000 souls. Today, Switzerland's population of 6.3 million precludes such a political format, but the arguments developed in *The Social Contract* are still operative, since the referendum continues to be the major mechanism through which Swiss society settles political disputes. Today, the tradition of the referendum may serve to delay the introduction into Swiss society of broader political conflicts, based on class divisions, which, in varying degrees, dominate the political structure and divisions of other European societies. The relative weakness of the Confederation compared with the high degree of social, political and economic autonomy of the Cantons makes the introduction of a uniform and universal medical care system an extremely difficult undertaking.

Health insurance is controlled by commercial companies and non-profit sickness funds with payments of premium paid by the insured. In 1955, a catastrophic illness insurance scheme was introduced to protect the 1000 small insurance companies and sickness funds. Unlike those in the U.S., the vast majority of the cantonal and communal hospitals and nursing homes are publicly owned, although they are funded through a mix of public and private sources. Today, health insurance has become a vital necessity for all Swiss citizens, and cantonal authorities are more important in the development of health policies than central government.

The taxation system that supports health care services, however, must be briefly reviewed to assess issues of justice. Approximately 55% of all taxes are collected by cantonal and communal authorities. Sales or consumption taxes produce some thirty-four percent of all government revenues. The bulk of non-sales taxes are collected from individual tax payers rather than corporations. Corporations have various ways of avoiding equitable taxation. ([2], pp. 48–71). For example, Ciba-Geigy accountants are able to declare on the balance sheet that the total value of a building is 1 SFR, a sort of symbolic value. Other company property, valued at several hundred million francs, is listed on the balance sheets at 15 million SFR. Any profits on capital gained abroad are not subject to taxation. Federalism, with the 27 different Cantons, accentuates tax inequality, e.g., on wealth of 50 000 SFR, Uri residents pay 7.5% whereas Zurich residents pay .8% in taxes. As Masnata-Rubattel (1978) remark, taxes do not serve to redistribute wealth in Switzerland ([42], pp. 137–140).

In summary, it is clear that taxation policy in Switzerland with its heavy emphasis on sales tax is one of the least progressive or equitable taxation systems in the industrial world.[8] While wealth, inheritance, capital gains, corporate income, and military exemption taxes all contribute to federal revenues, they do so from salary and wage earnings through both direct and indirect taxation. It is not, therefore, surprising that 10% of the Swiss population owns 90% of private wealth and less than 2% control two-thirds of the country's wealth ([30], p. 179). Switzerland spends even less per capita on both defense and health and welfare services than does the U.S. The defense expenditure by the federal government, cantons and communes together, does not exceed 9% of total expenditures and health and welfare spending by all those sectors of government represents only 17% of the total. The low expenditures on defense, compared with the U.S. and the U.K., reflect the fact that Switzerland has maintained a policy of neutrality for over 150 years and avoided involvement in both World Wars ([2], pp. 63–70).

What can one say about the adequacy of social and health care expenditures within and across the country? First, attempts to generate uniformity of service provision between cantons has been resisted throughout the history of Switzerland. Gilliand and Diserens have documented the lack of uniformity in health care access as well as differential morbidity and mortality rates between cantons [23]. Second, even with public ownership of hospitals and salaried hospital physicians, the system must rely on the multiplicity of small, inefficient insurance companies (Krankenkassen) to finance roughly 50% of hospital costs and most of physicians' services.

At the local level, health care services are in considerable disarray with many cantons exhibiting a surfeit of physicians and hospital beds. The federal government has turned its back on this problem, perhaps through a fear of attempting to impose uniform policies on the jealously guarded autonomy of the cantons. Over-supply of hospital beds and physicians has driven up health insurance rates until town meetings have been held to help explain the rising costs of health care. This situation is further exacerbated since health insurance for individuals is now a virtual necessity; hence, inflationary pressures are simply passed on in the form of higher insurance premiums. Federal subsidies to the health insurance schemes have not kept pace with the inflationary pressures, further increasing the cost component of health insurance to the individual, the Cantons, and the communes.

The circumstances in the City/Canton of Basel, with a population of approximately 250 000, illustrates these dilemmas. Today, 1/5th of the population is over 65, largely as a consequence of young families in the middle and upper income levels moving outside the city-canton. The city's tax base is reduced as a result. Notwithstanding the high proportion of aged persons, a recent study [58] estimated that the city had a surplus of 650–850 hospital beds associated with the largest cantonal physician-population ratio in the country. In the 1980s, Basel spent 1269 SFR ($808) per capita for health care, the highest expenditure by far of all cantons, and the true expenditure was probably double, if the additional expenditure of private individuals through their health insurance policies was taken into account. Hospital administrators and their medical staffs, when faced with declining revenues and a surplus of beds, are tempted and have apparently succumbed to the temptation to admit and maintain as many patients as possible. Many patients, particularly the elderly and the chronically ill, would probably benefit more from a health care system that emphasized home care services, sheltered accommodation, or even extended nursing home provision, rather than incarceration in facilities designed basically to treat acute illnesses. Attempts to redirect resources from tertiary care institutions to primary and home-based care are naturally resisted by the advocates of high technology medicine within the medical profession. Switzerland's problems, in terms of redistributing medical and social care resources, are made even more difficult by cantonal autonomy and the consequent lack of planning, coordination, and integration of its medical and social services. In this situation, even if the Geneva experiment (discussed earlier in this paper) is successful, its dissemination across the remaining 25 cantons will still remain problematic.

The care of patients with ESRD, in contrast with the care of and services available to the elderly, is relatively uniform across the country. While the rates of reimbursement for the elderly vary considerably from canton to canton, CLM pays a fixed sum of SFR 500 ($318) for each dialysis session to all programs [33]. When a rough adjustment for the higher standard of living of the Swiss compared with Americans is made, by using the figures for average per capita income ($15 000 to $11 000 respectively), it is apparent that the Swiss provide $174 as opposed to $150 per dialysis session, which is the highest reimbursement rate in the U.S. Clearly hospital administrators in Switzerland have a greater potential for utilizing excess income from ESRD to affect deficits incurred in other hospital operations than do their American counterparts, if they can indeed deliver services for less than $318 per dialysis session.

It is interesting to speculate about the origins of CLM, the reinsurance mechanism, in terms of its relatively early introduction (1955) and indeed, of its very existence. This reinsurance mechanism is, in effect, a form of national health insurance and seems to be antithetical to the philosophy of leaving the cantons and the communes with major responsibility for health care services. One of the driving forces must surely have been the economies of scale. In 1955, over 1140 private companies were offering health insurance in Switzerland. Although exact figures are unobtainable, it is said that some companies had fewer than 1000 members. Clearly one instance of catastrophic illness would bankrupt a small insurance company. This fact of economic life was, presumably, sufficient to eliminate any objections to an element of national health insurance on the part of the cantons. The year of introduction of the scheme, 1955, may be in part due to the post-war climate of opinion in Europe in which many countries had already introduced or were introducing policies in the provision of health services which, to varying degrees, protected individuals from the fiscal consequences of ill-health. The Swiss, perhaps, followed suit, albeit at a much lower level of development for nationally based indemnity schemes.

In general, it would seem that the Swiss ESRD program is less aggressive than the U.S. in that Swiss physicians recruit fewer younger and older patients than their American counterparts, producing a slightly lower overall ESRD treatment rate per million of population. Profit, in either the classic sense or in terms of increased salaries to physicians or other staff, plays no part in the Swiss system. However, hospitals do benefit if reimbursements for ESRD treatments exceed costs.

CONCLUSION

Our starting point, that a social system that operates with an individualist orientation is one that emphasizes the duties and responsibilities of individuals to provide for their medical care, whereas a social system that operates with a collectivist orientation is one that emphasizes the duties and responsibilities of the society at large to provide medical care to all, needs some refinement. Additional refinement and further elaboration of these contrasting statements are necessary in the case of both geriatric and renal care. The case of ESRD is somewhat more straightforward, largely because a single disease entity, although related to a number of treatment modalities, is a little easier to deal with than the more complex mix of health and illness states, social circumstances and various service provisions associated with the care and treatment of the elderly. We will, therefore, first discuss ESRD.

Our starting point for ESRD is 1972, a choice that excludes some interesting aspects of distributive justice in the experimental stage of a new and developing medical technology. Prior to 1972, selection or rejection of candidates for experimental dialysis therapy was dictated by strictly medical criteria, and except for their renal disease, patients needed to be relatively healthy. Since 1972, dialysis therapy and to a lesser extent renal transplantation, have become less problematic and, while not exactly routine procedures, they are certainly no longer experimental. In each of the three countries we reviewed, ESRD treatments are now accepted as part of the regular armamentarium of nephrologists and transplant surgeons.

Paradoxically, it is clear that in terms of distributive justice, Switzerland and the United States, whose societies and medical care delivery systems emphasize an individualistic orientation, provide the necessary resources so that all patients who can benefit from ESRD do so. By contrast, the United Kingdom, with a more collectivist orientation, provides insufficient resources to treat all cases of ESRD, as treatment facilities are limited and patients are denied care through age criteria that vary from Region to Region but generally operate at age 65.

The fundamental principle of the NHS – to provide comprehensive health care for everyone in the population, regardless of his wealth or social status – is clearly breached by the way in which ESRD patients are treated. But this is not the first exception within the NHS to the principles of equity and uniformity and, consequently, of distributive justice. At the inception of the NHS, 1948, inequity was institutional-ized through the decision to permit an admittedly small private medical care sector to survive. Moreover, the private sector survives, in large measure, by using hospital facilities, beds, operating theaters, nursing staffs and the like, in NHS hospitals to treat private patients. In 1974, private patients' payments for beds in the NHS were estimated at twelve million pounds ($18m), a sum that goes straight to the Treasury and is, presumably, reimbursed to the hospitals [3]. Using the figures quoted by Pincherle and Macini [51] for the costs of hospital and home dialysis and assuming no transplants, and using the estimate of 1700 untreated ESRD patients provided by Gabriel [20], a sum of approximately eighteen million pounds ($27m) would have been required in 1980 to treat these patients. By 1980 despite a slight reduction in the number of private NHS beds, it would seem that payments for such beds would have covered the additional expense involved in caring for the untreated ESRD cases. These sums are, of course, relatively small in relation to the overall 1984 NHS budget of £17 000 m, yet they loom large in terms of social justice. It would be easy to record at this stage a simple pejorative statement concerning the parsimony of the British and pass on to the next discussion, but this is insufficient.

The NHS is part of a much broader welfare state. As stressed earlier, while it is now understood that the welfare state plays a major role in the preservation of capitalist social structures (class divisions, inequitable distribution of scarce resources, preservation of the social and economic privileges of élites, etc.), it has nevertheless ameliorated some of the more extreme injustices of capitalist society. Food stamps are not necessary in the United Kingdom economy. The tax system is mildly progressive. Public housing for low income groups can represent any-thing from 30–50% of the housing stock in many cities and towns. No one runs the risk of pauperization through ill-health. Public transport by road or rail is reasonably efficient and affordable. Capital punishment is outlawed. Criminals are dealt with in a humane fashion, and drug addicts are treated as sick persons rather than as criminals. Although no effective solution has been found to compensate working class children for the lack of "intellectual and cultural capital" available in their homes compared with middle class children, and the consequent advantage this gives to the latter in the process of educational competition, working class children who are able to negotiate the examination hurdles face no financial barriers in all forms of higher education.

However, the retrenchment of the welfare state, which commenced with the Wilson government in the summer of 1975 and has continued under subsequent administrations, particularly the current and previous Thatcher governments, represents an attack on the basic premises of a caring society. For example, the availability of low-cost public housing is being reduced through the forced sale of council houses, a policy imposed on often reluctant local authorities by central government [25]. Nevertheless, there are theoretical and practical limits to the retrenchment of the welfare state, particularly in relation to medical care, since medical services contribute to all three essential components of societal survival – capital accumulation, the replacement of the work force, and the promotion of social harmony. The latter is particularly important if cut-backs in welfare provisions are further disruptive of social harmony and generate additional rioting against unemployment, urban blight and neglect, similar to that which occurred in the Toxteth Division of Liverpool in 1983.

Distributive justice and even the principles of equity and uniformity, therefore, become much broader issues than the allocation or non-allocation of resources for the treatment of a particular category of patients. Of course, the NHS is open to criticism because it ignores the needs of 1400–1700 ESRD patients, but it does not leave 11% of its population uninsured against the costs of ill-health as does the U.S., nor does it leave the bulk of health insurance costs to be borne by the less well-off segments of the society as does Switzerland. Of course the British government must be shamed into providing adequate funds for the treatment of all persons with ESRD and this can only be accomplished through the process of political agitation. Let us hope that a sufficient amount of political protest can be generated in the near future to correct this breach of the principles of equity and uniformity.

Both Switzerland and the U.S. have adopted a collectivist solution for the treatment of ESRD, with the Swiss achieving a somewhat more complete coverage of expenses for patients. In the U.S., especially in those states that do not operate a supplemental kidney program, middle class and "near poor" patients are faced with considerable out-of-pocket expenses. Even in those states with supplemental assistance programs, the energy required to sort out the bureaucratic morass of documentation and form-filling can be exhausting to the patient and his/her family [5].

After 1972, ESRD treatment in the U.S. became non-problematic in that no one who needed treatment for kidney failure was denied access.

In fact, this created a situation in which no criteria were operative concerning the selection of patients. Consequently, this absence of criteria eventually produced a new set of problems, initially apparent in the steep increase in the average age of ESRD patients. Patients with multiple system diseases and other conditions, which previously would have been perceived as contraindications, are now placed in ESRD programs despite the fact that in many instances quality of life is compromised. Moreover, this procedure really side-steps the need to make a clinical judgment: that those with the best medical chances of survival should have priority in access to treatment. This absence of criteria, in effect, contributes to an abrogation of the principle of distributive justice. In addition, the growth in the component of ESRD services is accomplished at great expense to the social system.

Current solutions to these dilemmas have generated attempts to produce rational and ethically sound procedures for solving the problems of microallocation. Leenen, for example, attempts to deal with these situations [39]. The dilemma, however, has a societal component that must be resolved in the long term. How much is any society prepared to spend on the next generation of medical technologies, heart transplants, artificial hearts, liver transplants, and the like? Again, it seems obvious that such issues can only be effectively resolved at the societal level using a collectivist orientation and procedures.

In the U.S. the principle of distributive justice is further abrogated by the obscenity of private profit-making in the ESRD program. From a collectivist perspective, profits are simply a waste of resources. The society-at-large is paying more for particular ESRD services than is required so that some physicians and stockholders can make a profit. Of the three countries, Switzerland, in terms of ESRD alone, achieves the most collectivist set of provisions, but it is funded through a taxation system that is the most regressive of the three.

It would be inappropriate to leave the discussion of ESRD without some comments on the benefits of these programs. While 23 000 of the ESRD patients in the U.S. in 1980 were described as 'ESRD disabled' and received disability benefits, an unknown[9] proportion of the remainder, including many of those with a successful transplant, were back at work, contributing to the economy, paying taxes and often higher premiums for health insurance coverage. Thus, the net cost of the ESRD treatment program was clearly less than $1.5 billion in 1980, but by how much no one can say.

The social benefits of ESRD are even more difficult to assess but no less important. What are the benefits to the family if grandfather or grandmother survives for a few more years or months through treatment of ESRD? Even to begin to assess this problem would require a lengthy analysis of the present state of family sociology, and we have neither the space nor sufficient expertise to attempt this here. Presumably this and other advantages of survival and/or increased longevity were sufficient to persuade Congress, L'Assemblée Fédérale, and to a lesser extent Parliament to provide funds for the treatment of ESRD.

In geriatric care it is only the British system that comes anywhere near realizing the principles of equity and uniformity and therefore meeting the overall criterion of distributive justice. Even in Britain the presence of private nursing homes and private medicine means that those with the ability to pay can avoid the inadequacies, perceived or actual, of the public sector. The public sector is financed on a progressive tax basis and forms the basis of the collectivist orientation.

Above all, distributive justice, equity, and uniformity are achieved through the procedure of public accountability which, in turn, is only manageable through the mechanism of a NHS, itself a component of the broader welfare state. To exercise collective responsibility a society must be in a position to plan, integrate, and coordinate all aspects of social and medical support. While the NHS does not meet this ideal, it and the other components of the welfare state are continually being improved and adjusted as new foci of concern emerge from the political process. Although it is difficult to achieve, obstetric beds in excess of capacity can be transferred to other needed medical services and, indeed, in some Regions this has been achieved with maternity wards being converted into geriatric facilities. The distribution of general practitioners across the country has been evened out. J. Tudor Hart's "inverse care law" [27] does operate, particularly in relation to hospital services, social work support, adequate provision in terms of both quality and quantity of low income housing, and so on, but all British residents have nearly equal access to a general practitioner, a considerable achievement in the 36 years of the NHS's existence.

Adequate geriatric care is, of course, only possible through a careful balance of medical and social care and it is in this area that the British system does least well. Local authority services are in part funded through block grants provided by central government and in part by taxes generated locally. The adequacy of local authority services are, therefore, dependent on the degree of commitment locally elected politicians are prepared to make to those social services that are the overall responsibility of the town or county council. Local elections in Britain reflect to a large extent the national division between right and left – the Labour, the Social Democratic, the Liberal and the Conservative parties. Consequently, although somewhat indirectly, local authorities are publicly accountable for the adequacy of their social provision and support services for the NHS. When the Conservative party is in power, distributive justice receives less attention since the philosophical basis of conservatism leans more toward individualism and the impersonal forces of the market place. A socialist government is, however, a likely alternative at each general election and left-wing parties dominate local politics in many cities and large towns. Hence, political forces concerned with issues of distributive justice are always present and provide an effective brake to policies that reflect excessive individualism. The trade unions provide a similar corrective to aggressive individualism in the economic sector.

In the U.S., by contrast, distributive justice is much more problematic. Attempts to plan medical services on a national basis have never taken root, and planning on a regional basis has recently been abandoned. Social service provision is subject to even less planning and coordination. Even the tax system operates to subvert distributive justice. Southern states, for example Texas, provide tax advantage to industry prepared to locate or relocate in their region. Texas has no income tax and Dallas has no city tax. (Presumably, taxes needed for state and city operations are generated through indirect taxes, which are, of course, regressive, as well as through property taxes which may inject a modicum of distributive justice.) In effect the "frost belt" and the older industrial states, with their higher rates of taxation, are subsidizing the economies of Texas and other southern states who use a variety of tax incentives to attract industry. The overall lack of distributive justice in the United States tax system is also, in part, responsible for the fiscal crises in cities like New York, Detroit, and Chicago. Migration of the middle and upper classes to the suburbs reduces the tax base of the inner city, thus imposing more fiscal pressure on the poor who remain.

Unless medical services are planned, administered, and financed on at least a regional basis and nationally for the more rarely needed and expensive tertiary services, the problems outlined above cannot be dealt with. In geriatric care this lack of planning and coordination is probably more serious than in other aspects of medicine. Effective care of the elderly requires cooperation between the primary, secondary, and tertiary medical care sectors, between medical and social services, between housing authorities, visiting nurses, home helps and finally backup provision of nursing homes. Such integration and coordination is difficult enough to achieve in Britain; in the U.S. it is practically impossible.

In Switzerland, the degree of confusion and possible duplication of expensive services through lack of planning, coordination, and integration is similar to that in the United States. Each canton has its own Minister of Health, its own public hospital or hospitals and social services, the latter often inadequate and underdeveloped. In a relatively small country of 6.3 million people, efficient use of resources is even more important if distributive justice is to be achieved. Duplication of facilities would represent an even larger component of the total resources allocated to medical and social care than might be the case in a larger country. In Switzerland's case the need for an equal and rational allocation of resources is paramount.

America, through high interest rates, exports its huge budget deficit to other Western nations by way of exchange rates and to third-world countries through higher interest rates charged against development loans, etc. In this way other Western nations and third-world countries are providing support to the U.S. economy and to the citizens of the U.S. In a sense, then, the sums made available to ESRD patients in the U.S. are partly funded by the citizens of other Western nations and some third-world countries. Clearly, the issue of distributive justice becomes even more complex when the perspective is enlarged to encompass the international scene and the world economy [17]. The European division of WHO is aware of this enormously complex problem and is beginning to explore the differing standards of medical care between the nations of the continent. WHO is also addressing the issue in its Alma Mater declaration, which calls for a minimal standard of health for all by the 21st century. Thus, from a discussion about the allocation of resources to the elderly and to ESRD patients, we reach the point where the issue of distributive justice has important implications for the whole human collectivity.[10]

School of Medicine
University of Missouri   *University of Maryland*
Columbia, Missouri   *Cantonville, Maryland*

NOTES

[1] The phrase 'rate of treatment' has been chosen to indicate the number of persons per million population actually receiving treatment for ESRD. It does not reflect either incidence or prevalence of renal disease.

[2] Roughly the equivalent, in terms of medical care administration, of States in the U.S. and Regions in the British NHS. This section partially relies on interviews by S. R. Ingman in 1983.

[3] Swiss per capita income was $15 000 in 1979 compared with $11 000 in the U.S.

[4] Partial payment for dialysis treatment and transplant surgery was required of ESRD patients through the named deductible and copayment requirements of Medicare. Such expenses were considerable and some states began to provide some or complete coverage of these out-of-pocket expenses. Further, for the small proportion of patients who opted for home dialysis the cost of their supplies was not included in the original legislation. These financial exigencies were largely removed by a further amendment to the Social Security Bill in 1978.

[5] Again, from a comparative perspective, it is worth noting that there is no equivalent of enterprises such as H & R Block in the United Kingdom. The U.K. income tax laws are such that few deductibles are possible and those there are, are sufficiently straightforward for individual taxpayers to handle. However, fringe benefits, avoidance of taxation, an income from capital and land and some other forms of tax avoidance are as common in the U.K. as they are in the U.S.

[6] These costs were calculated over 5 years and include training fees, physicians' fees, surgical procedures, equipment costs, annual supply costs, and average hospitalization costs. These figures should probably have applied to them an error factor of at least 10%, as anyone who has any familiarity with the problem of costing hospital and other medical procedures would readily agree.

[7] *Missouri Kidney Program* meets costs not usually covered by Medicare, e.g., transportation costs, home dialysis expenses, etc. *Unpaid* costs are presumably covered by institutions or from supplier resources.

[8] One might also note that the Swiss economy, particularly its banking, finance, and insurance sectors, benefits substantially from favourable liquidity ratios achieved by providing a 'secure' home for funds acquired by politicians and others in various African and Latin American countries by methods of acquisition – to put no finer point on it – which are clearly unethical. J. Saunier, *Le Pouvoir des Banques Suisses*, Édition Temps Actuels/La Vérité vraie, 1983. This is not to imply that it is only dishonest persons from the third world who make use of the unethical Swiss banking system and its favorable tax laws to non-national residents.

[9] While the European Dialysis & Transplant Association provides excellent data on the bio-medical aspects of dialysis and transplants in Europe, it provides little information of the socio-economic correlates of ESRD. In the U.S. the governing of information on ESRD is even more inadequate [52]. The difficulties associated with the collection of these data in the U.S. may be further exacerbated by the for-profit-sector in the ESRD treatment program. Private companies guard information as closely as other secrets of their production process and operations, and it may be that NMC is most reluctant to release data on its patient population, treatment costs, etc.

[10] This research was partially supported by a grant from the Research Council of the Graduate School, University of Missouri-Columbia.

BIBLIOGRAPHY

1. Ball, R.: 1981, 'Rethinking National Policy on Health Care for the Elderly', in A. Somers and D. Fabian (eds.), *The Geriatric Imperative*, Appleton-Century-Crofts, New York, pp. 21–38.
2. Boczek, B.: 1976, *World Tax Series: Taxation in Switzerland*, Commerce Clearing House, Chicago, Illinois.
3. 'Britain's Health Tangle': 1975, *The Economist* **257** (6896), 11–13.
4. Butler, R.: 1975, *Why Survive?: Being Old in America*, Harper & Row, New York, pp. 174–224.
5. Campbell, J. and Campbell, A.: 1978, 'Social & Economic Costs of ESRD: A Patient's Perspective', *New England Journal of Medicine* **229**, 386–392.
6. Cantor, M.: 1978, *The Divided Left*, Hill & Wang, New York.
7. Carboni, D.: 1982, *Geriatric Medicine in the United States & Great Britain*, Greenwood, Westport, Connecticut.
8. Carter-Jones, L.: 'Politics, Mortality & Economics – Are There Choices?' in F. Parsons & C. Ogg (eds.), *Renal Failure – Who Cares?*, MTP Press, Lancaster, England, pp. 99–106.
9. Coakley, D.: 1982, 'Introduction: The Elements of a Geriatric Service', in Coakley (ed.), *Establishing a Geriatric Service*, Croom Helm Ltd, London, pp. 9–16.
10. Cooper, M.: 1975, *Rationing Health Care*, Croom Helm Limited, London.
11. Department of Health and Human Services: 1983, 'New Regulations for Hospice Care', *Urban Life* **12**, 8.
12. Diamond, T.: 1983, 'Nursing Homes as Trouble', *Urban Life* **12**, 269–286.
13. Diamond, T.: 1984, 'Elements of Sociology for Nursing: Consideration on Care Giving and Capitalism', *Mid-American Review of Sociology* **9**, 3–22.
14. *Directory of Community Services for Boone County, Missouri*: 1984, Published by the City of Columbia, Office of Community Service (July).
15. Doyal, L.: 1983, 'Women, Health & Sexual Division of Labor: A Case Study of the Women's Health Movement in Britain', *International Journal of Health Service 13*, 372–387.
16. Eggers, P. Connerton, R. and McMullan, M.: 1984, 'The Medicare Experience with End-Stage Renal Disease: Trends in Incidence, Prevalence, and Survival', *Health Care Financing Review* **5**, 69–88.
17. Elling, R.: 1981, 'The Capitalist World-System and International Health', *International Journal of Health Services* **11**, 21–51.
18. Estes, C.: 1982, 'Austerity and Aging in The United States: 1980 and Beyond', *International Journal of Health Service* **12**, 573–584.
19. Freidson, E.: 1970, *Profession of Medicine*, Dodd Mead Co, New York.
20. Gabriel, R.: 1983, 'Chronic Renal Failure in the United Kingdom: Referral, Funding & Staffing', in F. Parsons & C. Ogg (eds.), *Renal Failure – Who Cares?*, MTP Press, Lancaster, England, pp. 34–40.

21. Gibson, D. and McMullan, M.: 1984, 'End-stage Renal Disease: A Profile of Facilities Furnishing Treatment', *Health Care Financing Review* **6**, 87.
22. Gill, D.: 1980, *The British National Health Service: A Sociologist's Perspective*, US-DHHS, NIH NO–80–2054, U.S. Government Printing Office, Washington, D.C.
23. Gilliand, P. and Diserens, M.: 1978, 'Santé Publique: Analyse factorielle des Disparités entre Cantons suisses', *Cahier de L'Institut Sandoz* **1**, 66–70.
24. Gough, I.: 1979, *The Political Economy of the Welfare State*, MacMillan, London.
25. Gough, I.: 1983, 'The Crisis of the British Welfare State', *International Journal of Health Services* **12**, 5–29.
26. Gutman, R.; Stead, W.; and Robinson, R.: 1981, 'Physical Activity & Employment Status of Patients on Maintenance Dialysis', *The New England Journal of Medicine* **304**, 309–313.
27. Hart, J.: 1971, 'The Inverse Care Law', *Lancet* **1**, 405–412.
28. Heller, T.: 1978, *Restructuring the Health Services*, Neale Watson Academic, New York.
29. Hobsbaum, E.: 1962, *The Age of Revolution 1789–1848*, New American Library, New York.
30. Hollinger, C.: 1974, *Die Reichen und die Superreichen der Schweiz*, Hoffman und Campe Verlag, Hamburg.
31. Ingman, S.: 1983, 'Geneva Geriatric Care System', *Missouri Gerontology Institute Newsletter* **4**, 4–5.
32. Ingman, S.: 1984, 'Nursing Home in Switzerland', *Missouri Gerontology Institute Newsletter* **5**, 4–5.
33. Ingman, S., Ingman, V., and Gloor, H.: 1984, 'Renal Care in Schaffhausen, Switzerland', Report prepared for Schaffhausen Cantonal Hospital.
34. Kayser-Jones, J.: 1979, 'Care of the Institutionalized Aged in Scotland and the United States: A Comparative Study', *Western Journal of Nursing Research* **1**, 190–200.
35. Kaplan, A.: 1981, 'Kidneys, Ethics, Politics: Policy Lessons of the ESRD Experience', *Journal of Health Politics, Policy & Law* **6**, 488–503.
36. Kutner, N.: 1982, 'Cost-Benefits Issues in U.S. National Health Legislation: The Case of the End-Stage Renal Disease Program', *Social Problems* **30**, 51–64.
37. Knolph, K.: 1983, 'Dialysis and Transplantation in the United States, and the Impact of Continuous Ambulatory Peritoneal Dialysis', in F. Parsons & C. Ogg (eds.) *Renal Care – Who Cares?*, MTP Press, Lancaster, England, pp. 78–88.
38. Kolata, G.: 1980, 'NMC Thrive Selling Dialysis', *Science* **208**, 379, 381–382.
39. Leenen, H.: 1979 'The Selection of Patients in the Event of a Scarcity of Medical Facilities – An Unavoidable Dilemma', *International Journal of Medicine & Law* **1**, 161–180.
40. Lens, S.: 1974, *The Labor Wars*, Anchor Books, New York.
41. Marmor, T. and Marmor, J.: 1970, *The Politics of Medicare*, Routledge & Kegan Paul, London, England.
42. Masnata-Rubattel, C. et F.: 1978, *Le Pouvoir Suisse: Séduction Démocratique et Répression Suave*, Christian Bourgois, Paris.
43. Morabia, A.: 1983, *Médecine et Socialisme: Politiques sanitaires en Suisse et dans les sociétés capitalistes avancées*, Édition d'en Bas, Lausanne.

44. Murray, R.: 1955, *Red Scare: A Study in National Hysteria 1919–20*, McGraw-Hill, New York.
45. Navarro, V.: 1982, 'The Labor Process and Health: A Historical Materialist Interpretation', *International Journal of Health Service* **12**, 5–29.
46. Navarro, V.: 1984, 'Selected Myths Guiding the Reagan Administration's Health Politics', *International Journal of Health Services* **14**, 321–328.
47. O'Connor, J.: 1973, *The Fiscal Crisis of the State*, St. Martin's Press, New York.
48. Offe, C.: 1972, 'Advanced Capitalism & the Welfare State', *Politics and Society* **2**, 479–88.
49. Offe, C.: 1972, 'Political Authority & Class Structure: An Analysis of Late Capitalist Societies', *International Journal of Sociology* **2**, 73–108.
50. Parsons, R.: 1983, 'Five Years Since Stirling-A Process Review', in F. Parsons and C. Ogg (eds.), *Renal Failure – Who Cares?*, MTP Press, Lancaster, England, especially pp. 5–16.
51. Pincherle, G. and Mancine, P.: 1983, The Costs of Treating Chronic Renal Failure', in F. Parsons and C. Ogg (eds.), *Renal Failure – Who Cares?*, MTP Press, Lancaster, England, pp. 25–34.
52. Prottas. J.; Segal, M.; and Sapolsky, H.: 1983, 'Cross-National Differences in Dialysis Rates', *Health Care Financing Review* **4**, 91–103.
53. Pryor, J.: 1983, 'Comparison of Facilities in the United Kingdom and in Europe for Dialysis & Transplantation', in F. Parsons & C. Ogg, (eds.), *Renal Failure – Who Cares?*, MTP Press, Lancaster, England, pp. 17–24.
54. Relman, A. and Rennie, D.: 1982, 'Treatment of End-Stage Renal Disease: Free but not Equal', *New England Journal of Medicine* **303**, 996–998.
55. U.S. President's Commission for the Study of Ethical Problems in Medicine and Biomedical and Behavioral Research: 1983, Report on *Securing Access to Health Care*, Vol. I, Table 9.
56. Taylor, D.: 1984, *Understanding the NHS in the 1980's*, Office of Health Services, London, England.
57. Taylor, T.; Aitchison, J.; Parker, L.; and Moore, M.: 1975, 'Individual Differences in Selecting Patients for Regular Haemodialysis', *British Medical Journal* **2**, 380–381.
58. Thalmann, V.: 1983, 'Nicht nur Basel-Stadt hat zu viele Spitalbetten', *Schweizerische Krankenkassenzeitung* **16**, 250–251.
59. Trafford, A.: 1982, 'Doctor's Dilemma: Treat or Let Die', *U.S. News & World Report*, 53–56.
60. Vladeck, B.: 1980, *Unloving Care: The Nursing Home Tragedy*, Basic Books, New York.
61. Vladeck, B.: 1984, 'Medicare Hospital Payment by Diagnosis-Related Groups', *Annals of Internal Medicine* **100**, 576–591.
62. Wautburg, W.: 1983, 'Kostenentwicklung im Gesundheitswesen – Meinungen und Tatsachen,' *Schweizerische Aerztezeitung* **64**, 761–768.
63. Wauters, J.; Spertini, F.; and Parsons, V.: 1983, 'Selection Criteria and Physician Bias in the Treatment of End-Stage Renal Failure, Results of a National Survey', presented at 4th Congress of International Society for Artificial Organs, Kyoto, Japan.
64. Wauters, J. *et al.*: 1983 'Regionalized Self-Care Haemodialysis', *Journal of American Medical Association* **250**, 59–62.

H. TRISTRAM ENGELHARDT, JR.

THE BAD, THE UGLY, AND THE UNFORTUNATE

There are a number of things wrong with the world, only some of which can be remedied. There is much wrong due to moral evil, due to the culpable actions of persons. In principle, such could be avoided. If men and women would renounce their evil ways, a moral millennium would dawn. In any event, those who act immorally can be punished and, as far as possible, forced to make restitution. After all of this, much in the world would still be undesirable and unpleasant. Much will be found ugly and unfortunate even if no one is to blame. Given the limitations of humans, it is not possible in principle to set aside all that is disvalued. The moral millennium may require divine grace, *pace* Pelagius, but to set the unfortunate and ugly aside will take a miraculous intervention into the physical structure of the world. Still, if all problems cannot be solved, it does not follow that some of that which is ugly and unfortunate should not be overcome.

Norman Daniels [1] and Stanley Ingman with his colleagues, Derek Gill and James Campbell [5], have addressed ways in which health care systems attempt to blunt the unfortunate outcomes of disease, disability, and aging. They have provided us with proposals for somewhat similar, yet still in important ways different, patterns for allocating resources to the elderly. These proposals require the redistribution of resources and income. Daniels attempts to justify a redistribution to health care (1) that will improve the chances of reaching a normal life span, (2) that is dedicated to extending life expectancy or life span only once a normal life span or expectancy is generally secured for all (or most?) ([1], p. 24), and (3) that favors the provision of personal health care and social support services over the use of exotic or expensive health extending services in old age. Daniels argues that we should approach these issues not as concerns of justice understood as a fair solution to a dispute between competing groups, but as a prudent investment of resources by individuals who, as they age, will pass through various stages of life. Daniels has given us an engaging account of justice as prudence. Ingman, Gill, and Campbell address the issue of distributive justice through the case example of providing hemodialysis

263

Stuart F. Spicker, Stanley R. Ingman, and Ian R. Lawson (eds.),
Ethical Dimensions of Geriatric Care, 263–270.
© *1987 by D. Reidel Publishing Company.*

to the elderly. Ingman *et al.* see justice in health care as requiring uniformity and equality in health care such that the elderly may receive at least some exotic or expensive treatment, which they would not receive under Daniels' account of justice as prudence.

Daniels' account of justice to the aged as the expression of disinterested individual choice regarding the investment of resources across stages of life offers a view that, within certain qualifications, can endorse the British limits on the allocation of hemodialysis for the elderly, a policy criticized by Ingman *et al.* Daniels is open to endorsing limitations on the extension of life of the aged when such extension would involve a movement of resources away from areas disinterested individuals would prudently choose, to areas they would not choose. One could thus imagine not providing dialysis for severely senile individuals. Ingman *et al.* do not critically address the issue of whether equality and uniformity are in all cases prudent goals. Daniels has provided the basis for a major criticism of positions such as Ingman *et al.*

The issues I will raise go to the roots of both papers insofar as they presuppose a general redistribution of resources on behalf of the realization of patterns of allocation, which are purportedly discovered to be morally obligatory under a view of justice. To employ force in order to achieve such goals requires being able: (1) to discover successfully the morally correct pattern for distribution, and (2) to demonstrate authority for its imposition. In addition, even if such authority is available and such a pattern can be discovered, (3) there will be circumstances when the imposition of the appropriate pattern by force will prove to be imprudent [2]. Difficulties in meeting the first two conditions spring from the problem of discovering a morally authoritative concrete view of the good life. Such a view presupposes a particular hierarchy of goods and values. However, to discover which particular hierarchy is canonical, one will need to make a moral judgment. That is, one must determine which ranking of benefits and harms is authoritative. Such a determination requires a moral sense. One cannot appeal to consequences, since the significance of consequences can itself be determined only by a particular ranking of benefits and harms, which is what is in question. One needs a moral sense of which ranking is appropriate. To select the correct moral sense, one will need to appeal to a higher level moral sense ad indefinitum. Appeals to disinterested observers will not help. If the observer is truly disinterested, it will not be able to decide between competing hierarchies of values and harms, between competing moral senses. If the disinterested observer is able to decide, in any concrete fashion, that is, beyond simply indicating what sorts of actions are consonant with the notion of resolving issues through mutual respect, not force, one will have begged the question and have imputed to the disinterested observer a particular moral sense. If one cannot discover a particular pattern for distribution as morally correct, how will one then derive the authority to impose a particular pattern? If an authoritative pattern cannot be derived from a sound moral argument, it will need to be created through the mutual consent of those involved [2]. It is this difficulty of discovering a concrete view of the moral life that leads liberal democracies to give such a central place to free and informed consent. They do so not simply because of liberal sentiments, but because of the difficulty in establishing with authority what individuals ought to do. When one cannot discover what individuals ought to do, one can at least ask them what they will agree to do.

The moral limits on collectivist solutions to the unfortunate outcomes of the natural and social lotteries spring from limits on the plausibility of arguments to establish as morally canonical a particular pattern for the distribution of resources as well as from limits on the plausibility of arguments for the use of force in achieving particular worthy goals such as the provision of health care for the elderly. One may be forced to tolerate individualistic solutions not because of the value one places on individualism, but because of limits on the plausibility of particular views of just distributions or of claims to have the authority to use force in realizing them. As the justification for particular patterns of distribution and/or their imposition by force weakens, one then moves to allocation of resources that depends on the free consent of those involved.

It is because of difficulty in establishing the authority to use force to impose a particular pattern of allocations on the unconsenting innocent that individual rights to forbearance are more easily established than welfare rights. It is easier to impeach the authority to use force in imposing a particular view of the good life on unconsenting innocents than it is to establish the right to use such force and thus to fashion systems of coercive beneficence. Skepticism supports a public policy that is procedural in relying on peaceable negotiation, with the participants retaining rights of privacy, dissent, and withdrawal unless such rights are individually relinquished. All-encompassing mandatory national health systems become morally suspect. The more comprehensive the intrusions into individual lives, the more well justified such intrusions must be and therefore the more difficult such justification becomes.

The more one concludes that it is implausible that all property, resources and services are wholly either in societal or private hands, the more one will come to embrace mixed capitalist systems. The mixed capitalist accounts criticized by Ingman *et al.* may be morally unavoidable, so that one will need to focus not simply on what is a desirable goal for the community to achieve, but also on what the society *may* achieve with coercive force. The discussions developed by Daniels, Ingman, *et al.* are more plausible, the more they are seen in the restricted context of exploring what ought to be done with societal resources, recognizing that individuals will still possess private resources that set a moral limit on the taxing authority of the state and its coercive power to constrain the services of individuals. Insofar as truly private resources exist, they provide a basis for acting independently of, and contrary to, the ways in which a society has decided to allocate its common resources. So far as private resources exist, protected by autonomy rights (which individual autonomy rights may simply be a shorthand for the plausible limits of state authority), one may then act privately to buy around the system. Indeed, one might be required morally to acknowledge the black market, including the black market for health care services such as the provision of additional or supplementary services forbidden under a comprehensive national health service or the charging of fees in violation of an all-encompassing price freeze, prospective payment system, or other all-embracing restraint on the free exchange of services for money. In many cases the black market is one of the last refuges of freedom and one of the least irradicable tributes to liberty. It reminds us that respect of freedom leads to inequality and tragedy, and perhaps in certain circumstances to disorganization and inefficiency. However, it should be noted, black markets are usually very efficient. The constraints on freedom that elicit a black market can be unavoided only if one avoids what Robert Nozick has described as "ownership of the people, by the people and for the people" ([6], p. 290). Individual rights provide moral limits to communities achieving morally important goals.

If, for the purpose of discussion, one presumes that all resources are neither societal nor private, and that individuals have a right (even if such rights only signal the limits of collective authority) to use the resources left over to them after taxation to purchase the services of others, one will then morally have to accept a two-tiered health care systems as one finds throughout most of the world. In fact, the suggestion that one should move toward a more encompassing health care system than currently exists in the United States may raise issues of justice in taxation that are not often discussed. For example, it has been proposed that the very wealthy in the United States be taxed 28% on every dollar and that at death the highest marginal taxation on their estates be 50%. This means that on portions of their estates some individuals would be able to dispose freely of only 36% of what they have earned. Depending on the circumstances of the acquisition of such funds, it may be very difficult to show that the community owns 64% of what one produces. Indeed, there are grounds for being skeptical about the community's claim to have the authority to tax away more than 50% of any dollar earned, even if the taxing is done in stages. After all, it is the force user, the tax collection agency, which must justify its use of force, not the individual who non-coercively and peaceably comes into possession of a fortune. A not implausible final maximum amount of taxation would be a fifty–fifty split. One should remember, it is individuals and corporations that create the wealth claimed by states. Further, in 1982 those with incomes over $20 000 paid over 84% of individual federal income taxes, and those with incomes over $100 000 a year provided over 17 percent of federal income tax revenues from individuals, though less than 1% of the tax returns fell in that category.[1] Despite what Ingman et al. state, those with incomes over $100 000 are making a major contribution to the American federal budget. It is clearly not the case that they are paying no taxes. It is difficult to see how Ingman et al. can correctly characterize the U.S. tax system as only mildly progressive (indeed, at best one might describe the United States tax system as not unrestrainedly confiscatory). In any event, even if one concluded that the present U.S. levels of taxation could be morally justified, one would have learned from this exercise by seeing that the onus of justification is on the shoulders of the government. The government must justify its right to confiscate any portion of what individuals produce, since the state is the force user.

The development of modern methods of securing profits from medicine may raise a question of whether special taxes or special constraints on profit-making are appropriate. However, one must first ask why profit from health care should be considered or taxed any differently than profit from other endeavors. If individuals or corporations direct their energies to making profits from the practice of medicine or from for-profit hospitals, one may be able to argue that they should be exposed to the same level of taxation as other individuals or corporations. But the feeling on the part of some, or many, that profit from such undertakings is morally suspect does not in itself justify higher taxes on such endeavors, much less proscribing profit from them. It is difficult to see why it is less distasteful to make a profit from the practice of law, than from the practice of medicine. Both provide important and needed services. Beyond that, the profit motive tends to direct the best energies and to reward efficiency and ability. It may not be a mere accident that capitalist countries have tended to produce more wealth and better health care systems for their populace than have Marxist systems. The U.S.S.R., for instance, has one of the worst health care systems in the industrialized world ([3], [4]). In addition, inequalities in access to health care have been recognized in more affluent Marxist health-care systems [7].

On the other hand, Switzerland has less inflation, less unemployment, and more health care benefits for more of its citizens then the United States, United Kingdom, or the U.S.S.R., though according to Ingman, Gill, and Campbell, Switzerland has a tax system less progressive than the United States and provides a "worse" distribution of wealth. The criticisms that Ingman *et al.* lodge against Switzerland's taxation system must seriously be considered in the opposite spirit as a plausible recipe for insuring better health and greater wealth for all. Perhaps the Swiss have recognized that taxes on corporations are inefficient modes of revenue collection since such costs are passed on to customers and to the citizens of the countries in which the corporations operate. Among other things, if taxed too highly, corporations will locate elsewhere and take the jobs and capital they attract with them. Despite what Ingman *et al.* say about profit, one must remember that profit is what individuals, corporations, or governments offer to individuals or corporations to attract their free and uncoerced cooperation in the realization of selected goals. The alternative to offering a profit is the acquisition of services through coercion, through appeals to charitable contributions, or through relying on accidental areas of collaboration. The fact that individuals and corporations can sell their products at prices greater than their costs is a major motive force of capitalism's productivity. It is by appeals to profit that things usually get done in the absence of coercion. In this light, it may not at all be paradoxical, as Ingman *et al.* suggest, that "in terms of distributive justice, Switzerland and the

United States, whose societies and medical care systems emphasize individualistic orientation, provide the necessary resources so that patients who can benefit from ESRD do so" ([5], p. 40).

Utilitarian considerations of the productivity of the market may set substantial prudential limits on the taxing authority of the state, apart from the limits one may need to recognize as a matter of principle. Even if one does not acknowledge deontologically justified rights of individuals to sell their health care services and of patients to buy them privately, one might still argue that the concurrent existence of a private health care tier provides a useful criterion against which a public health care system can be measured. If individuals can privately purchase hemodialysis or other services denied to patients who must rely on the public health care system, there will be a goad to public discussion of the proper range of health care that should be available through communal mechanisms. Thus, utilitarian considerations and arguments on principle favor two tiers of health care provision to individuals in general, including geriatric populations: one communally supported and the other supported by private funds.

Open and critical discussion of the amount of health care to be provided for the elderly is essential to a democratic society, especially if one concludes that the proper amount to be provided cannot be discovered through arguments such as those of Norman Daniels. The more one becomes skeptical of such arguments, the more one will need to rely on common agreement. The creation of a particular system of health care services for the elderly can be understood as a society creating for itself a particular level of health care insurance against losses at the natural and social lotteries. No particular level of health care insurance against the natural and social lotteries will be morally mandatory. Different societies with different views of how to come to terms with risks will fashion insurance policies with different scopes and costs. Some societies, such as Texas, may offer fewer social welfare benefits in order to attract corporations and jobs. In addition, particular groups of private individuals may, and very likely will, acquire supplementary insurance. This is, in fact, what occurs in the United States when individuals purchase insurance to supplement their Medicare coverage.

In summary, there will not be a way to discover answers to the important question of the extent to which we are morally obliged to provide funds for health care, geriatric or otherwise. The proposals of Norman Daniels fail because he hopes to be able to discover an answer where an answer can only be created. The proposals by Ingman, Gill, and Campbell must be disallowed, however attractive, in that it is implausible that the state or society has authority to realize encompassing goals of equality through the use of coercive state force. Criticisms of both of the papers thus rest on skepticism regarding the capacities of reason. The more skeptical one is about the ability to discover the proper distribution of resources or to establish the authority to impose that distribution by force, the more one must acquiesce in the heterogeneity of health care systems and the different levels of care that will likely result from the free choices of individuals and democratic societies. Though philosophy may not be able to give us a final vision of the good life or authority for its realization through coercive state force, it can still help us understand the limits of state power and the nature of respectful negotiation among peaceable individuals.

Center for Ethics, Medicine and Public Issues
Baylor College of Medicine
Houston, Texas

NOTE

[1] The very few individuals who earn more than a million dollars per year paid 2.4% of all individual income tax in 1982. More than $1 out of 50 was collected from them alone. 'Tricklenomics', *The Wall Street Journal* (April 11, 1984), 28. However, only 8408 individuals had such incomes. Leonard Wiener and Robert J. Morse, 'The Swelling Ranks of U.S. Millionaires', *U.S. News and World Report* (March 18, 1985), 53. One might conclude that the super-rich are forced to shoulder an undue burden of payment.

BIBLIOGRAPHY

1. Daniels, N.: 1987, 'Equal Opportunity, Justice, and Health Care for the Elderly: A Prudential Account', in this volume, pp. 197–221.
2. Engelhardt, H.: 1986, *The Foundations of Bioethics*, Oxford University Press, New York.
3. Feshbach, M.: 1980, U.S. Department of Commerce, Bureau of the Census, in *Rising Infant Mortality in the U.S.S.R. in the 1970's*, Series P–95, no. 74.
4. Feshbach, M.: 1982, 'Between the Lines of the 1979 Soviet Consensus', *Problems of Communism* **31**, 27–37.
5. Ingman, S., D. Gill, and J. Campbell: 1987, 'ESRD and the Elderly: Cross-National Perspectives on Distributive Justice', in this volume, pp. 223–262.
6. Nozick, R.: 1974, *Anarchy, State, and Utopia*, Basic Books, New York.
7. Szalai, F.: 1986, 'Inequalities in Access to Health Care in Hungary', *Social Science and Medicine* **22**, 135–140.

EPILOGUE

THE EIGHTH STAGE OF HUMANITY

IAN R. LAWSON

ELDERLY DEPENDENCY AND SYSTEMS FAILURE:
OBSTACLES TO A PROSTHETIC SOCIETY

I. INTRODUCTION

Almost 30 years ago, in Scotland, I became immersed in the deep end of
elderly morbidity, practicing in the newly-contrived geriatric services of
the regional hospital systems. Capital resources – the supply of long-
term care beds in particular – were limited. Nevertheless, the physicians
being salaried and the hospital services being prospectively budgeted,
we had absolute clinical and managerial control over what was available
as a free service to the public and the elderly. There was no "third
party" system of mediating the care supplied, and an entire absence of
utilization review activities. The principal role of the senior geriatricians
was to make the limited hospital and long-term beds serve the region's
thousands of disabled and dependent elderly.

Making these resources work led to some innovative practices in
rehabilitation, day-hospital care, and in pre-admission assessment of the
elderly person at home prior to institutionalization. The last was made
not by a social worker or nurse, but by the senior physician in overall
charge. It was not a friendly house call; it was a rigorous assessment of
diagnoses, needs, and priorities; it was a brokering of conditions and
services to families under great stress; it was a gatekeeping function to
large scale services, themselves under stress.

This practice was operationally critical and intellectually illuminating.
It provided an ecologic perspective of the relations between elderly
persons and their social and physical environment [7], a phenomeno-
logic revelation like Claude Bernard's physiology of *le milieu intérieur*.

Improving these relations (by what one might term 'prosthetic re-
arrangements') was as crucial to the operation of the geriatric services as
managing the illnesses and disability by medication and rehabilitation.
The overall success rate was a return to the community of about one in
two, or better, of seriously disabled patients [2] – patients who had been
consigned by the rest of the medical care system for permanent institutional care.

273

Stuart F. Spicker, Stanley R. Ingman, and Ian R. Lawson (eds.),
Ethical Dimensions of Geriatric Care, 273–290.
© *1987 by D. Reidel Publishing Company.*

In the United States, I still practice domiciliary visits, somewhat against the counter incentives of money and time. In a more rational system, this practice may yet be found to be the ultimate technology of geriatric care. Paradoxically, when I display at case level what these hard-driven Scottish services were able to provide, in comprehensiveness of assessment by physician specialists in the home, and in rehabilitation (like taking a year to rehabilitate complexly disabled stroke patients), I am likely to be told, "We cannot afford that here." Yet it should be emphasized that these practices were developed out of circumstances of extreme parsimony relative to the United States.

Of these peculiar British methods, this ecologic deployment of clinical medicine in *macro-systems* of geriatric care, I think there is about as much comprehension by the major agencies now as there was in the 1970s of the economics of E. F. Schumacher's "Small is Beautiful," or the "just in time" inventory control and "quality circle" systems of the Japanese. And as our automobile industry was then, so are our medical institutions now: wealthy, hardware oriented, massively redundant (in parts), supremely confident about the future, yet operationally unprepared for it. Medicine has yet to experience the "pain of Detroit" – the distress experienced by the auto industry toward the imperatives of the oil crisis, the small car, and Japanese competition. Quick though we have been to import the hardware of Britain – the Charnley hip replacement and the CAT scanner – our institutions and fiduciary agencies balk at the software imports to meet the "geriatric imperative": the revamping of fiscal and institutional systems, and the cognitive-perceptual developments of medical role implied by my opening remarks.

The reasons are not apathy, ignorance, or ineptitude, but cultural, firstly, and organizational, secondly. They result from a defective biosociology combined with a high-level denial syndrome and systems perseveration in the for-profit mode.

II. THE BIOSOCIOLOGY WE SHOULD ORGANIZE ABOUT

A. *Dependency* in the elderly refers to a set of consequences that arise from disease, that nevertheless require to be conceptualized apart from disease, but still require to be managed integrally with disease. Organ system impairment has intruded on personal life and function to cause disability. When the nature and extent of the disability are such as to require the ongoing support of other persons for the performance of the essential activities of life, there is a state of dependency. Essential activities include not only the vital daily activities of eating, drinking, breathing, and excreting, but a hierarchy of other activities – grooming, moving, prehensile function, expression, communication, symbolic functions, senses of sovereignty and fulfillment, and meaningful relations with others. States of dependency are therefore composites of distresses. They include experiencing chronic or recurrent symptoms in a long life lived hitherto largely without them; the severe restriction on personal options; the loss of a hopeful future; a self-conscious dependency on a physical environment in which one was once the unthinking master; and a reluctant, tedious dependency on those to whom one was formerly parent, spouse, or peer.

Dependency in the elderly is significantly different from dependency in the younger disabled person, to an extent that requires different management and supporting systems. There is the underlay of biologic senescence (as opposed to growth and vitality); the multiplicity of diseases present (as opposed to a single disabling cause); the inferior adaptability of the elderly brain (so called "plasticity") both toward its own damage and in compensatory activity toward other organ systems; the difficult inversion of the parent-child role when middle-aged children begin to give care to aged parents; and the problems of providing care when, as a spouse, one is close in age and frailty to the dependent person.

B. *Dementia* is a special and epidemic cause of dependency. Theories about competitive marketing, choices for the healthy, prudent consumer, coupon usage and other devices for best-buy, etc., are all very well as responses to a political and administrative pre-occupation with the same. But consider how they "fit" with the "consumer" who has dementia.

The sense of where one is in time and space is confused, and even objects are unstable: "Where is my watch? I put it there and now it's not. . . . " The identity of the physical environment becomes ambiguous and mistakeable: "This is the room I was born in, I think," said an elderly lady just admitted to a nursing home because of her dementia.

All serious disability involves an altered relationship to the environment as can be seen in the approach that disabled persons adopt toward it. Watch a neurologically or arthritically impaired person scan a room across which to walk and sit – like a mountain climber confronting nature, looking for niches and toe holds. In dementia, however, our abilities to circumvent, to scan for the opportune, to exploit, and to accommodate to our environment are fundamentally impaired just when other disabilities mandate it. Increasingly, the demented person relies on an accommodating environment, one that will actively respond to him or her in a prosthetic way. Observation of relatives of such people will show that they develop, consciously or unconsciously, supportive practices and attitudes in a kind of prosthetic symbiosis.

III. FISCAL IMPEDIMENTS TO ORGANIZATIONAL DEVELOPMENT

Medical care toward this complex set of needs, which includes the needs of the frequently stressed supporter as well as the dependent person, is markedly deficient ([4], [10]). Its deficiencies result from a "diagnosis-related," fiscal mediation of services that excludes from Medicare funding any broader support of disability and the dependency situation [8].

A. *Rehabilitation* is the next critical component of the geriatric system, after an adequate professional assessment of person and milieu. It should work at all levels of the functional hierarchy. In particular, and beyond mere technique, by a milieu of optimism and group care, motivation and morale is "imploded" into the dependent and disabled by the healthy therapists – another example of prosthetic symbiosis. Indeed, because the rehabilitation process is such a complex of symbiotic transfers between damaged and undamaged brains, it is difficult to break it down categorically (as the U.S. fee-for-service requires it to be) into itemized actions and units of time.

"Third party" conditions require proof of probable benefit before services are supplied, and, by a casuistry around the concept of "skilled services," limit rehabilitation to the provision of intense packages of "therapy" applied for short duration, weeks at most and rarely months. But the nature of the disability in the elderly is such that the principal way to discriminate who are the likely beneficiaries is by allowing rehabilitation to most if not all. Especially in the early phases of illness with disability, rehabilitation needs to be applied in low intensity but for long duration. No wonder our long-term care bed needs are burgeoning and our rehabilitation services for elderly atrophying under present cost-containing conditions.

In contrast, the United Kingdom geriatric services deployed rehabilitation *en masse*, pragmatically and (judging from the containment of the long-term care beddage) effectively. It was the operational outcome – minimizing the consignment of elderly to the limited long-term care beds – that justified the rehabilitative means, whatever they were. In the United States, *the outcome is of indirect interest* to the financial and institutional systems underpinning the fee-for-service providers. The accounting systems demand and receive tedious detail of rationale applied to the *process*, not the *outcome* of care. The means are not justified by the ends but by the means. In the case of the elderly, there is also an inherent systems competitiveness between the administration of Title XVIII, which largely supports (or denies) rehabilitative care, and Title XIX, which supports the majority of elderly who gravitate into long-term care for the lack of it [9].

B. *Day care* for the elderly and respite care (to relieve stressed care givers) probably represent the supreme antitheses of geriatrics, USA *vs.* UK. One of the earliest studies demonstrating the conditional responsiveness of brain-impaired elderly was conducted by Cosin and his colleagues, in 1958, in Oxford [5]. In a therapeutic milieu some improvement of function did occur which, however, lasted only as long as the input. In our society, with its demands for "onward and upward" in rehabilitation, such care is termed "maintenance" and is routinely denied. Most day care and respite care are unsupported by medical insurance, whereas in the UK they became important adjunctive services to the in-patient geriatric services. Unabashedly prosthetic, day care and respite care were designed to support rather than improve on present function, to keep things going for family and patient as long as possible. In the UK, we reckoned that we could not make our limited resources work without such prosthetic adventurism. Given the institutional redundancy and plethora of resources of the USA, we continue to wonder whether it, and measures like it, are cost effective.

IV. DESIGNING THE PROSTHETIC SOCIETY

One of our earlier titles of the Symposium from which this volume developed was "Aging and Dependency in a Prosthetic Society," for that was our view of the 21st century and the imagined context. We foresaw a need for prosthetic systems as well as technology, systems that would eventually be designed around the "3D" dimensions of the biology of elderly need: disease(s), disability, and dependency. Systems friendly to the elderly dependent person would be characterized by the following:

A. Access Engineering

"We'll make it to you if you can't make it to us";

- house calls by doctors as well as nurses;
- inclined ramps to doctors' offices;
- receptionists who will organize the transport as well as schedule the visit;
- reminder-systems for elderly who forget;
- follow-up and outreach in regard to those who don't and can't make it unaided to the system because of the "Inverse Care Law"[1]

Access engineering makes a good test case in primary care of geriatric prosthetic systems vs. traditional practice. To the colleague who says, "What is geriatrics? I look after old people too", one can commonly respond: "You look after old people who can meet three criteria that operate in your current office practice:

(1) they must make it to your office or other site of your choosing;
(2) they should give their story within the time allotted by your scheduling, which is a dividend of your expected income over volume of visits and procedures needed to support that income; and
(3) in the case of 50% of practitioners, elderly must pay at the time service is rendered, on Medicare non-assignment; thus they must also sustain the cash-flow burden, the administration of their submission of claims, and the large deficit between what the physician charges and what the 'third party' reimburses.

Elderly who fail in any of these criteria will tend to appear no longer in your office, leaving you with a sense that you fill all older persons' needs, since that is exclusively the population your system has selected for you."

B. Informatics

"We do the remembering for you. Wherever you turn up, however you turn up, in our medical care system what we need to know about you will be there with you."

The system takes the time to get the whole story right the first time (medical history, hospitalizations, family network, medications, prior adverse reactions, etc.), never loses or "forgets" it, and has the essential profile available wherever and whenever that older person materializes in the office or emergency room, at home or by phone, or a thousand miles away visiting a daughter and gone ill.

C. Ecologic Sensitivities and Priorities

"Keeping people caring for people." This is not a slogan for a facile, unsystematic volunteerism, but an operational goal for provider institutions and fiscal agencies: targeting the needs of the supportive complex that maintain the elderly person – not abstract categories called "eligible diseases." (The DRG system organizes this abstraction further into profit-making-and-losing diagnoses.)

D. Systems Interconnectedness and Continuity

"Any way you come at us, wherever you turn up, we're ready and able to cope." This will involve a necessary degree of redundancy organized into the care system, in order to cope with the unstable, somewhat unpredictable profiles of elderly needs. At the present time, redundancy exists in the most expensive sectors of care only: the in-patient beds and emergency rooms of the general hospitals; which is why, until the DRG system, they played profitable catch-all for the inadequacies of the rest of the system – the lack of psycho-geriatric in-patient care, of day care and respite care, and of physician and nursing coverage for the more intensively ill in nursing homes and at home. The turn of the fiscal screw via DRGs may, on the short term, look economic. (As a nation we are, after all, addicted to the "quarterly return" to the exclusion of longer term financial and industrial forethought; so why not with medical and social services with the elderly?) The effect on the system is to remove the expansion joint in an otherwise rigid structure of provision. For, concurrent with DRGs, there has been no liberalizing of benefits for home care and skilled nursing facilities in order to cope with elderly now extruded from the formerly relaxed phases of hospital and its emergency entry services.

E. Enhancing the Patient's Life Options (Social and Recreational)

Despite disability and organ-system impairment, what unambiguous benefits a spirited staff of nurses, rehabilitation and recreational therapists can achieve in this top-level tier of human function! And at what modest costs relative to the overpriced and oversold commodities of, say, the cardio-vascular purveyors whose returns on saving life turn out to be quite modest, and redolent with ethical conundrums.

F. Re-deploying Medical Care Ecologically and Systems Interactively

The organization of medicine into office and hospital based specialties, each focusing on separate organ systems, has old roots and rationality. It also is highly profitable and enjoys parallel support and organization in Academe.

Unfortunately, the phenomenon of elderly disability and dependency spills over the boundaries of such circumscribed professional interests and also beyond the walls of their exercise. Commonly, multiple organ systems have gone synchronously awry, and there are also the critical interactions with environment already described. NASA recognizes the need for a general systems engineer when it builds a space satellite: such an engineer has responsibility for seeing that the design and provision of components is compatible with the viability and function of the satellite as a whole, in a specific and highly comprehended environment. Modern medicine sentimentalizes about the need for a "generalist" role, but in the enormously expensive area of critical care medicine one sees the daily abnegation of it. Such is the general dissolution of overall responsibility, that the first principle enunciated by one eminent University department of medicine, in regard to withholding further resuscitative care, was to "identify the responsible physician" [3]; this, after some days or weeks of what one might call poly-specialist care. But if the modern geriatrician is to exercise some "trans-systems" role, it will not be at case level only. Whole person concerns lead inevitably to a wish to reshape professional and institutional systems, as in the successful

British case. This systems engineering role is not only an unfamiliar one to most physicians in the United States but, with the rise of the administratively rather than the clinically qualified manager, will now be hotly contested.

Nevertheless, the agenda of the 1985 annual meeting of the Gerontologic Society of America showed a remarkable growth of geriatric assessment units – in-patient and out-patient. They are "natural" to the V.A. system as they were to the British Health Service, where there is (a) contractual responsibility for a defined population, (b) a financial disinterest by (salaried) medical providers in the provision of care, and (c) some integration of long-term care with acute care. They are also experimental and fiscally dubious add-ons to the civilian hospital system for the converse of reasons stated.

G. Good Systems Should Promote Provider Creativity, Empathy, Altruism, Advocacy, and Stamina.

This is less an ethical statement than a set of operational objectives. The manufacturing mega-systems of America were put on the defensive by the "quality is loyalty" ethos propagated by the Japanese and behaviorally engineered through their small-systems team work. We plead overmuch that their culture, etc., makes them inapplicable, as we plead similar impediments to the British geriatric system.

Meanwhile, the "corporatization" of American health care on the profit motive and the continued attempt to ration the particulars of individual care through nationwide regulations are dual and complementary trends of extraordinary counter-productiveness when related to the variables and pathos of elderly need. But it is also what they do, in the long term, in the social conditioning of professional workers that must be viewed as antithetical to the idea of autonomy and conscience in professionalism and to its exercise of social leadership and innovation. Belatedly, it is only since these two confluent influences – corporate and governmental regulation – have affected the finances of hospitals and doctors that alarmed debate has been engendered in the medical press. The nursing home, home care, and the nursing and allied health professions have been longer exposed to them, often with doctors coopted into a compliant rather than a protesting role because they had perceived these activities as peripheral to their own professional and

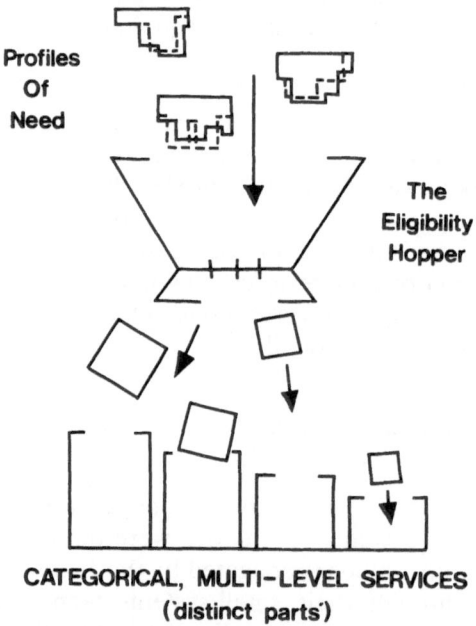

CATEGORICAL, MULTI-LEVEL SERVICES
('distinct parts')

Fiscal-mechanical model. While the eligibility hopper may be operated by
professionals, they act as certifiers and sorters, applying fiscal-regulative criteria (the
hopper grid). Profiles of elderly need must be summated into one of three or four fiscal
quanta; then placed accordingly in predetermined, "distinct part" levels. Reductionism
of the biology of need and fixity of the separate service components are implicit.
[Used with permission.]

economic interests. Yet, it was only a question of time before they
applied universally [8]. Less and less do health professionals, even those
without direct financial gain, mediate the quantities of services. More
and more are they operators of fiscal rules: they work (what I call) the
"eligibility hopper" [9].

V. UNFRIENDLY SYSTEMS, OR TOO MUCH OF WHAT WE HAVE
NOW

The institutional, administrative, and, to a degree, professional systems

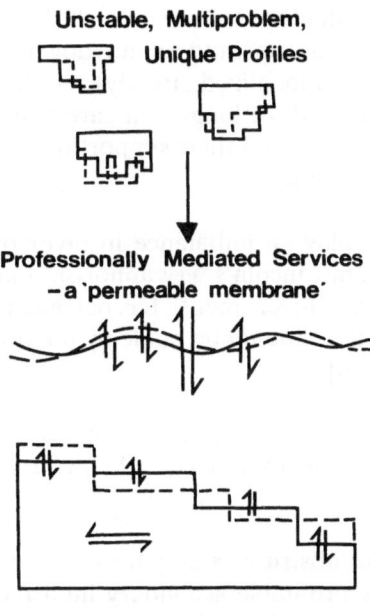

Unstable, Multiproblem, Unique Profiles

Professionally Mediated Services
– a `permeable membrane`

PLASTICITY & MOBILITY OF
MULTILEVEL SERVICES

Cybernetic-biologic model. Functionally autonomous professionals operate as a permeable membrane does with regard to the cell – coping with different configurations of need (molecules), engaged in matching needs in both admissions and discharges (active transport), and able to affect also the internal structure and character of the agency or institution (the cell), as well as the extramural community in need.
[Used with permission.]

are unfriendly if the characteristics I have listed as "friendly" are accepted as criterial but absent.

A. In the creation of Medicare and Medicaid there was an unproven premise that the elderly would be overusing of services, and that relatives would be opportunistic and abandon responsibility. Hence in Medicare, fiscal barriers were created like "deducts" and "co-pays."

Inspections are made, in regard to Medicaid, for surreptitious divestments of assets to accelerate elderly persons' "spend down." Only when spend down is accomplished will Medicaid services of "alternatives to institutional care" be deployed; but since all discretionary income has been lost with loss of assets (and spend down has more commonly occurred by being institutionalized already) the situation is irreversible. Yet the accelerated growth of long-term care beds is still attributed to the behavior of the elderly and their supporting relatives rather than to the systems' maladaptation.

B. These systems display an imbalance in favor of vendor profits and convenience. Abraham Lincoln's wry aphorism that "everyone likes to make a buck out of the Government" has become a principle of modern Government: megabucks *should* be made by the private sector or things won't happen right [10].

C. These systems offer losing choices in the vital-for-life tier of functions: I have been asked by elderly, "Do you want me to take the medication or to eat?"

D. These systems are abstruse, ambiguous, and capricious, as seen by the disabled user. A profitable secondary industry has arisen, indeed, around the consumer's perplexity in the face of the *designed* complexity of Medicare insurance. Advertisements of "Medigap" insurance imply security and comprehensiveness when the terms of Title XVIII (as refined by what a Federal judge once called "the disentitling practices of the Social Security Administration") make insecurity inherent and inescapable.

E. The provider services and agencies supported by these fiscal systems are a mosaic of services (with gaps and overlays) arranged around what is profitable to supply. Non-profit agencies supplying unprofitable but necessary services lose their margin of support when for-profit agencies "cream it off."

F. They communicate confusingly and redundantly: The phrase – "This is not a bill" – at the top of a computer-generated statement may end with a bottom line implying an unpaid balance for which someone else *is* going to bill. They convey incidental rather than essential information.

Coast to coast, in whatever hospital one lands as an elderly person (and our mobile 80-year-olds have about a one-in-five chance of that in any one year), the status of one's Medicare deductible will follow one to the penny; but not the list of drugs one is taking, or prior adverse reactions, or prior medical care experience.

G. Overall, these systems are ominous, arbitrarily discontinuous, and morale defeating. They display *perseveration* – a symptom of brain damage which implies a persistence in activity which is no longer appropriate. They emanate from public policy and legislation and as such carry more profound messages to the elderly than if they were purely the design of private enterprise. Contemporary elderly may sooner or later infer that, whatever the conventional ethic in which they were raised, it is now everyone for himself: society has reneged, in a tortuous and unspoken way, on the social contract of an earlier day. Our contemporary care systems dismay the beneficiaries as much by the conditions of their capricious benevolence as by the costliness of what they tender. Affecting older people at a time when other losses and deprivations are mounting, they engender disillusionment and a self-defensive advocacy, which, carried into voting practice, could affect the whole adversely.

H. When I address my middle-aged peers in medicine, I attempt to stimulate them with the phrase – "Most of us here today will become 80 or 90 years old, so let's do better by the elderly now so that we shall be better done by in our advanced agedness." But I do believe that in the minds of many there is the thought – "I can buy myself out of what the rest may suffer."

This thought is not un-American. Indeed, it is very American. It is upheld by many elderly who think they have prudently secured themselves against the vagaries of the future. It also explains why geriatric medicine has not prospered as a practice specialty and why a prosthetic society is unlikely. Put simply, disability with dependency is not one of the scenarios that either the medical profession or the public envisages for itself. It is an unacceptable antithesis to individualism, to "can do" entrepreneurialism, and to the faith in the technologic fix. We have, in some four generations or so, become acculturated and deconditioned to disease and debility as inevitable realities of life.

VI. ELDERLY DEPENDENCY IN THE TWENTY-FIRST CENTURY

When men and women died prematurely, at thirty or forty years of age, a century or more ago, they had endured premonitions of mortality for years prior to their deaths. As in deprived societies today, the killers of mankind then were also wasters of mankind: malaria, tuberculosis, rheumatic heart disease and nutritional deficiency states. One not only died young, one had little of the experience of feeling well.

An average 70-year-old of today enjoys a state of health and freedom from symptoms that most 40-year-olds did not experience a century or two ago. While it is a uniquely opportune time for our sector of mankind, it is also what makes entrance into debility and disability so hard to comprehend: for unlike our forefathers, we are not used to feeling ill or impaired. Serious illness of that sort comes as an unwelcome surprise and it appears that we are not able to envisage or design for our own worst case.

A respondent, Mr. Joseph T. Engelberger[2], has reacted to my clinical cases by describing prosthetic hardware that could be designed around some of the disabilities I described: (1) the electronic companion – a CRT sit-in which would be voice-activated by the tremulous anxiety of the elderly dementing person left for a time on his or her own, which would play video responses by relatives according to her questions (in designing for dementia, there is at least one advantage in stereotypic repetitiveness); (2) a prosthetic brain – something of myself at fifty, say, captured in silica, to remind me and "cue" me at 80 with regard to essential life activities, and (3) robotic home-helps, etc. All quite feasible, but presently without a market. Why?

Mr. Engelberger recently told me that his friends, like my medical friends in middle age, are thinking of their retirement communities as having marinas and golf courses, but not (save the thought) being equipped prosthetically. The private part of my geriatric practice is conducted in a condominium complex for pre- and post-retired people of educated, upper-middle-class origins. After ten or more years of existence, over one in four of its population is now over 75 years. Indeed, it has golf courses and swimming pools, but its housing has no grab bars in the bathrooms. When I described this to an audience of largely out-and-about elderly, they laughed at my counterpoint; for them, grab bars were not yet essentials of safety and survival. We and they rest on a facile optimism concerning the power of the individual over his or her own future – save enough, jog enough, exercise enough, occupy our minds, take vitamins and eat a balanced diet; if that fails, we hope that modern medicine will provide the technologic fix. The years I have experienced with well elderly and disabled elderly convince me that few can envisage or sympathize with the condition of disability or dependency. The final attribute of a prosthetic society must therefore be prosthetic advocacy. Who should provide it? I say the doctors!

At the present time, five percent of the Medicare budget goes to support renal dialysis for less than a half percent of its beneficiaries. Truly there was advocacy by dying patients and their families. But there was also, and still is, an effective advocacy by doctors whose careers, whose income, and whose self-respect all depend on the prosperity of that kind of care we call renal dialysis. We argue about the cross-national ways and means to support it, attributing to the United Kingdom a rationing of care not practiced here. In fact, over about the same period when renal dialysis became funded by Medicare, there has occurred a stripping of nursing home and home care benefits under Medicare, but by attritional processes too diffuse to engender public or professional reaction. Likewise, influenza vaccination of the elderly is still unfunded by Medicare. Meanwhile (at this writing) the Secretary of the Department of Health and Human Services is reviewing the merits of Medicare support of heart transplantation. Supported by the most glorified and wealthiest of medical lobbies, the cardiovascular specialties, one has little doubt about the outcome – or that its funding will affect adversely the less glamorous provision of care elsewhere, care where no doctors stand to lose by its diminishment.

As a member of the late Governor Ella Grasso's Connecticut Nursing Home Commission, I was impressed by the skew of the representation at our public meetings: nursing home administrators and owners, relatives, lawyers, employees, etc., but (as I recollect) not one patient, not one doctor, and not one hospital administrator was present.

VII. CONCLUSION

The disabled and dependent elderly now comprise about 1½ million in long-term care institutions (whose beddage far exceeds that of hospitals) and three or four times that at home, mostly in the care of very elderly

spouses and middle-aged children. The numbers will grow exponentially
as the 21st century approaches. To provide for the elderly, both hu-
manely and economically, their protean needs must be directly
addressed. Systems and professional behavior can and must be remod-
elled about that growing and irreducible biologic reality. Given the
counter-productiveness of many of our present arrangements (attempting to make the reality fit the systems), doing better by the elderly may
not cost more, *if* there is a preparedness to reapportion resources.

But there's the rub. The present situation is not fortuitous. There is,
as Vladeck [11] describes representatively from the field of nursing
home care, a complex etiology. There has been massive public funding,
to great private and professional profit, of care of the elderly. Hence
there are strong vested interests (by no means confined to nursing home
operators) in preserving a profitable status quo, even if it is eminently
archaic and overtaken by the demography of dependency.

The academic perspective, such as ours, is not necessarily helpful.
Having observed the entanglements of fiduciary and entrepreneurial
interests in the huge field of nursing home care, Vladeck concluded that
the best we could hope for was a regulated form of minimal adequacy
[11]. As in its 1977 report, the Institute of Medicine continues to profess
uncertainty about the role of geriatrics in medicine; indeed, in its
America's Aging: Health in an Older Society it is hardly mentioned at
all. [1] Knowing more, studying situations and conditions, is still the
Institute's priority; seeking causes reflects that Academe has its own
vested interests too – the funding of research.

When we do not really wish or know how to act, we plead the need for
certainty and the inadequacy of sound data on which to proceed (such as
funding modestly educated women to act as home helps). When we
really want to act (as in cardiovascular surgery), we plead that there is
adequate empiric data available and that probabilistic thinking in medicine makes us never certain anyway. We are Sophists about our preferred actions and inactions.

To make better provision of the care of today's dependent elderly, we
indeed possess adequate information. There is still the problem of
translating *knowing about* (of which we have many professors) to
knowing how (for which we have few skilled practitioners). There is the
matter of no longer paying ourselves well for the profitable things we
prefer to do and paying ourselves less for the presently unprofitable
things that we ought to do. The economics, like the systems we devise

(we should come to perceive) actually reflect rather than significantly determine our moral choices.

School of Medicine
University of Connecticut
Farmington Connecticut
  and
Professional Building
Heritage Village
Southbury, Connecticut

NOTES

[1] "The availability of good medical care tends to vary inversely with the need for it in the population served. This inverse care law operates more completely when medical care is most exposed to market forces and less so where such exposure is reduced" [6].
[2] Mr. Engelberger is President, Transitions Research Corp., Danbury, Connecticut.

BIBLIOGRAPHY

1. *America's Aging: Health in an Older Society*: 1985, Report of the Committee on an Aging Society, Institute of Medicine and National Research Council, National Academy Press, Washington, D.C.
2. Arnold, J. and Exton-Smith, A. N.: 1962, 'The Geriatric Department and the Community – Value of Hospital Treatment in the Elderly', *Lancet* **2**, 551–553.
3. Committee on Policy for DNR Decisions, Yale New Haven Hospital: 1983, 'Report on Do Not Resuscitate Decisions', *Connecticut Medicine* **47**, 477–483.
4. Comptroller General Report to the Congress of the United States: 1979, 'Entering a Nursing Home – Costly Implications for Medicaid and the Elderly', U.S. General Accounting Office, PAD–80–12.
5. Cosin, L. Z., Mort, M., Post, F., Westrop, C., and Williams, M.: 1958, 'Experimental Treatment of Persistent Senile Confusion', *International Journal of Social Psychiatry*, **4**, 24–42.
6. Hart, J. T.: 1971, 'The Inverse Care Law', *Lancet* **1**, 407–412.
7. Lawson, I. R.: 1969, 'Confusion in the House: the Assessment of Disorientation for the Familiar in the Home', *Psychiatric Quarterly*, **43**, 225–239.
8. Lawson, I. R.: 1974, 'Professional Standards Review Organization and Care of the Elderly', *Journal of the American Medical Association* **229**, 311–313.
9. Lawson, I. R.: 1977, 'The Antitheses between Fiscal and Clinical Systems in Geriatric Care', in E. J. Hinman (ed.), *Advanced Medical Systems: The Third Century*, Symposia Specialists, Miami, Florida.

10. Office of Technology Assessment: 1984, 'Technology and Aging in America', Congress of the United States, OTA-BA-265, Summary, 28–32.
11. Vladeck, B. C.: 1980, *Unloving Care: The Nursing Home Tragedy*, Basic Books Inc., New York, N. Y.

NOTES ON CONTRIBUTORS

Margaret P. Battin, Ph.D., is Associate Professor of Philosophy, Department of Philosophy, University of Utah, Salt Lake City, Utah.

Tom Beauchamp, Ph.D., is Professor of Philosophy, Department of Philosophy, and Senior Research Fellow, Kennedy Institute of Ethics, Center for Bioethics, Georgetown University, Washington, D.C.

Richard W. Besdine, M.D., is Associate Professor of Community Medicine and Health Care, School of Medicine, and Director of The Travelers Center on Aging, University of Connecticut Health Center, Farmington, Connecticut.

Baruch A. Brody, Ph.D., is Leon Jaworski Professor of Biomedical Ethics and Director, Center for Ethics, Medicine, and Public Issues, Baylor College of Medicine, Houston, Texas.

Jacob A. Brody, M.D., is Professor and Dean, School of Public Health, University of Illinois, Chicago, Illinois.

James D. Campbell, Ph.D., is Assistant Professor of Behavioral Science (sociology), Department of Family and Community Medicine, University of Missouri at Columbia, Columbia, Missouri.

Teo Forcht Dagi, M.D., M.P.H., M.T.S (F.A.C.S.), is a neurosurgeon, Neurosurgery Service, Walter Reed Army Medical Center; Adjunct Professor of Law, Georgetown University Law Center, and Senior Research Fellow, Kennedy Institute of Ethics, Georgetown University, Washington, D.C.

Norman Daniels, Ph.D., is Professor of Philosophy and Chairman, Department of Philosophy, Tufts University, Medford, Massachusetts.

Nancy Neveloff Dubler, LL.B., is Director, Division of Legal and Ethical Issues in Health Care, Department of Epidemiology and Social Medicine, Montefiore Hospital and Medical Center, The Bronx, New York.

H. Tristram Engelhardt, Jr., Ph.D., M.D., is Professor of Medicine and of Community Medicine, and Member, Center for Ethics, Medicine, and Public Issues, Baylor College of Medicine, Houston, Texas.

Molly Gavin, M.S.W., is Regional Director, Connecticut Community Care, Inc., Bristol, Connecticut.

Derek Gill, Ph.D., is Professor and Chairman, Department of Sociology in Behavioral Science, University of Maryland, Catonville, Maryland.

Thomas Halper, Ph.D., is Professor and Chairman, Department of Political Science, Baruch College and the Graduate School, City University of New York, New York.

Daniel M. Hausman, Ph.D., is Associate Professor of Philosophy, Department of History and Philosophy, Carnegie-Mellon University, Pittsburgh, Pennsylvania.

Joseph M. Healey, Jr., J.D., is Associate Professor of Community Medicine and Health Care (health law), School of Medicine, University of Connecticut Health Center, Farmington, Connecticut.

Stanley R. Ingman, Ph.D., is Associate Professor of Behavioral Science (sociology), Department of Family and Community Medicine, University of Missouri at Columbia, Columbia, Missouri.

Gayle Kataja, R.N., B.S.N., is Nurse Case Manager, Connecticut Community Care, Inc., Bristol, Connecticut.

Ian R. Lawson, M.D., F.R.C.P (Edin.), F.A.C.P., is Clinical Professor of Community Medicine and Health Care (geriatric medicine), School of Medicine, University of Connecticut Health Center, Farmington, Connecticut.

Richard A. Lusky, Ph.D., is Assistant Professor of Community Medicine and Health Care (sociology), School of Medicine, University of Connecticut Health Center, Farmington, Connecticut.

Randolph M. Nesse, M.D., is Associate Professor of Psychiatry, Department of Psychiatry, University of Michigan, University Hospital, Ann Arbor, Michigan.

Albert Rosenfeld is Adjunct Professor, Division of Human Genetics, University of Texas Medical Branch, Galveston, Texas, and Consultant on Future Programs, March of Dimes Birth Defects Foundation, White Plains, New York.

Stuart F. Spicker, Ph.D., is Professor of Community Medicine and Health Care (philosophy), Division of Humanistic Studies in Medicine, School of Medicine, University of Connecticut Health Center, Farmington, Connecticut.

INDEX

The Philosophy and Medicine Book Series

Editors

H. Tristram Engelhardt, Jr. and Stuart F. Spicker